Trade and Environment

Trade and Environment: Difficult Policy Choices at the Interface

Edited by
Shahrukh Rafi Khan

Zed Books
LONDON • NEW YORK

in association with
The Sustainable Development
Policy Institute
ISLAMABAD

Trade and Environment: Difficult Policy Choices at the Interface was first published by Zed Books Ltd, 7 Cynthia Street, London N1 9JF, UK and Room 400, 175 Fifth Avenue, New York, NY 10010, USA in 2002 in association with the Sustainable Development Policy Institute, PO Box 2342, Islamabad, Pakistan.

Distributed exclusively in the USA by St Martin's Press, Inc., 175 Fifth Avenue, New York, NY 10010, USA

Cover designed by Andrew Corbett
Set in Monotype Dante by Ewan Smith, London
Printed and bound in the United Kingdom by Biddles Ltd, Guildford and King's Lynn

A catalogue record for this book is available from the British Library.

Library of Congress CIP data has been applied for.

ISBN 1 84277 098 5 cased
ISBN 1 84277 099 3 limp

Contents

Tables and Boxes

Tables

Boxes

Contributors

Dr Tariq Banuri, the founding member and former executive director of the SDPI, has a PhD in economics from Harvard University. His past positions include environmental policy adviser, IUCN-Pakistan (1991–92), visiting lecturer, Harvard University (1990), and assistant professor UMass/Amherst (1984–88). He is a member of several editorial advisory boards of international academic journals, and governing boards of NGOs. He is the author of several papers and publications: editor, *Economic Liberalisation: No Panacea* (Oxford, 1990); co-editor, *Financial Openness and National Autonomy* (Oxford, 1992); co-editor, *Governance for Sustainable Development* (IIED, 1992); editor and main author, *Pakistan's National Report to UNCED 1992*; co-editor, *Who Will Save the Forests?* (Zed, 1993); co-editor, *People, the Environment and Responsibility* (Parthenon, 1995; co-editor, *Defining and Operationalising Sustainable Human Development* (UNDP, 1994). At present, Dr Banuri is senior research director, SEI/Tellus, Boston, USA.

Aaron Cosbey is an associate and a senior adviser on trade and investment at the IISD. He has published widely on the subject of trade and sustainable development. He serves on Canada's Environmental Sectoral Advisory Group on International Trade and its Academic Advisory Council on International Trade, and as an alternate board member of the International Centre for Trade and Sustainable Development. He holds a master's degree in Development Economics from Dalhousie University, Canada.

Mark Halle is director of the IISD's trade and environment work and directs IISD's European Office in Geneva. He has worked for the Conference on Security and Cooperation in Europe, for UNEP and for WWF-International, where he served as acting deputy director-general. He was for many years with the IUCN, where he was most recently director of global policy and partnerships. He also serves as chairman of the board of the International Centre for Trade and Sustainable Development. He holds a postgraduate degree in historical studies from the University of Cambridge, UK.

Abdul Matin Khan, senior research analyst on the Technology Transfer for Sustainable Industrial Development (TTSID) project, joined the SDPI in 1996. He earned his BSc and MSc in Chemical Technology from the Institute of Chemical Engineering and Technology, Punjab University. He specializes in production and process controls in the chemical industries. He is working for the identification of environmental problems of six industrial sectors, suggesting pollution-control measures, developing training material for their technical staff and holding training sessions. Before joining the SDPI Mr Khan held senior appointments in the corporate industrial sector and in plants that involved chemical engineering. He conducted studies of process evaluation, inventory control and other technical aspects in the chemical industries. He is a member of the Institute of Chemical Engineers, Pakistan.

Haroon Ayub Khan, visiting fellow at the SDPI, has a master's degree in international management from the Graduate School of International Studies, University of Denver. He has worked in various capacities with the UNHCR, UNICEF and the UNDP in Peshawar (Pakistan), Baghdad (Iraq), Dadaab (Kenya) and New York (USA). His direct NGO association includes a tenure as the deputy country representative for Oxfam, Afghanistan and three years at the SDPI. As an SDPI visiting fellow, Mr Khan is currently working on a regional initiative on environment and non-traditional security.

Shahrukh Rafi Khan has a PhD in economics from the University of Michigan. He was formerly a research economist and head of the Public Policy Section at the Pakistan Institute of Development Economics. He has taught economics at the State University of New York and at Vassar College. He has worked at the SDPI since 1993 and is currently the executive director. He has done academic consulting for many agencies, including the East–West Center, the World Bank, UNESCO and USAID / International Food Policy Research Institute, Norad, the Asia Foundation, the Human Development Foundation, North America and UNICEF. In addition to numerous journal articles he has edited or co-edited several books and monographs. These include *Higher Education and Employment Opportunities in Pakistan* (UNESCO, 1988), *Just Development* (Oxford, 1997), and *Fifty Years of Pakistan's Economy* (Oxford, 1999). He has written several books, including *Profit and Loss Sharing* (Oxford, 1987), *Government, NGOs and Communities in Social Sector Delivery: A Study of Collective Action* (Ashgate, 1999), *Do IMF and World Bank Policies Work?* (Macmillan, 1999), and *Reforming Pakistan's Political Economy* (Vanguard, 1999).

Mahmood A. Khwaja, a research fellow at the SDPI, earned his PhD from La Trobe University of Science and Technology, Melbourne, Australia. He is involved in the Institute's project on Technology Transfer for Sustainable Industrial Development (TTSID) and is responsible for the Environmental Monitoring Programme for Industry. Dr Khwaja has over sixty publications to his credit, including 'Mercury pollution in Ghana coastal commercial fish', *Environment Tech Letters*, 10 (1989); 'Environmental Pollution', EPA, NWFP (1991); 'Utilisation of wastes from chemical and galvanised industry (zinc chemical)', *Pakistan Journal of Scientific and Industrial Research* (1993); 'Pollution due to effluents from tanneries and leather industries in NWFP', *Pakistan Journal of Scientific and Industrial Research* (1995); 'Initial environmental examination (IEE) of CFC for metal and metal based industries', Savant ED Pakistan (1995); 'Nitrate and nitrite contamination of sub-surface water in NWFP', *Journal of Chemical Society of Pakistan* (1998). He was the founding editor of *Chemistry and Industry* (1995), and associate editor, Proceedings of First National Chemistry Conference, University of Peshawar (1989).

Konrad von Moltke is a senior fellow at the IISD, at the WWF, Washington, DC, and at the Institute for International Environmental Governance at Dartmouth College. He is a visiting professor at the Free University, Amsterdam. He was founding director of the Institute for European Environmental Policy. He holds a doctorate in medieval history from the University of Göttingen, Germany.

Professor Adil Najam, a visiting fellow at the SDPI, is assistant professor of international relations and environmental policy at Boston University. He is also associate director of the MIT-Harvard Public Disputes Programme. He has graduate degrees in technology and policy and in civil and environmental engineering from MIT. His research interests include North–South environmental negotiation, policies for climate change, sustainability indicators, assessment and monitoring, the role of NGOs in environmental policy and multilateral environmental agreements (MEAs). He was part of the team that wrote the Pakistan National Conservation Strategy (PNCS). He has also taught at the Massachusetts Institute of Technology (MIT), the University of Massachusetts at Boston and the School for International Training (SIT). He is currently a lead author in Working Group III (Social and Economic Issues) of the IPCC. He has worked as a consultant with a number of international agencies and NGOs, including the IUCN, the IDRC, CIVICUS and UNITAR. He co-edited (with Lawrence Susskind and William Moomaw) two volumes of 'Papers on International Environmental Negotiation'. His forthcoming book is titled *Global Environmental Negotiation in an*

Unequal World: A Strategy for the South (Zed Books). His scholarly research has appeared in journals such as *Environmental Conservation*, *International Environmental Affairs*, *Population Research and Policy Review*, *International Studies*, *Nonprofit Management and Leadership*, *Nonprofit and Voluntary Sector Quarterly*, *Climatic Change*, *Journal of Environment and Development* and *Development Policy Review*.

Victoria Kellett is a project manager at the International IISD, based in Canada. She specializes in international sustainable development with a particular focus on the politics and economics of the emerging international climate change regime. She is the network coordinator for the Climate Change Knowledge Network, which brings together 14 organizations from developing and developed countries to conduct research and action on climate change policy. She holds an MA in international affairs from the Norman Paterson School of International Affairs, Carleton University, Ottawa, Canada and a BA in political studies and Russian from the University of Auckland, New Zealand.

Abbreviations and Acronyms

ADBP	Agricultural Development Bank of Pakistan
AIJ	activities implemented jointly
APTMA	All Pakistan Textile Mills Association
APTPMA	All Pakistan Texile Processing Mills Association
ATC	Agreement on Textiles and Clothing
BOD	biological oxygen demand
CCRI	Central Cotton Research Institute
CDM	clean development mechanism
CER	certified emissions reduction
CFC	chlorofluorocarbon
CITES	Convention on International Trade in Endangered Species
COD	chemical oxygen demand
CTE	Committee on Trade and Environment
DPG	domestically prohibited goods
EM	environmental management
ESC	Environmental Standards Committee
EPA	Environmental Protection Agency
ETPI	Environmental Technology Programme for Industry
FAO	Food and Agriculture Organization
FAP	Farmer Associates Pakistan
FCCC	Framework Convention on Climate Change
FPCCI	Federation of Pakistan Chambers of Commerce and Industry
GATS	General Agreement of Trade in Services
GATT	General Agreement on Tariffs and Trade
GMO	genetically modified organism
ICM	integrated crop management
ICSID	International Centre for Settlement of Investment Disputes
ICTSD	International Centre on Trade and Sustainable Development
IDRC	International Development Research Centre

IFOAM	International Federation of Organic Agricultural Movements
IISD	International Institute for Sustainable Development
IPCC	Intergovernmental Panel on Climate Change
IPM	integrated pest management
IPNS	integrated plant nutrient system
LDC	least developed country
MAI	Multilateral Agreement on Investment
MEA	multilateral environmental agreement
MFA	Multi-Fibre Arrangement
MFN	most favoured nation
MIGA	Multilateral Investment Guarantee Agency
MNC	multinational corporation
M&R	monitoring and reporting
NAFTA	North American Free Trade Agreement
NEQS	National Environmental Quality Standards
NFC	National Fertiliser Corporation (Pakistan)
NGO	non-governmental organization
OCAC	Oil Companies Advisory Committee
OECD	Organisation for Economic Co-operation and Development
OICCI	Overseas Investors Chamber of Commerce and Industry
PAPA	Pakistan Agricultural Pesticides Association
PASSCO	Pakistan Agriculture Seed and Storage Corporation
PCSIR	Pakistan Council for Scientific and Industrial Research
PEPC	Pakistan Environmental Protection Council
PETF	Provincial Environmental Trust Fund
PPM	process and production method
PSDF	Provincial Sustainable Development Fund
QELRO	Quantified Emissions Limitation and Reduction Objective
SD	sustainable development
SDPI	Sustainable Development Policy Institute
SPS	sanitary and phytosanitary measures
TBT	technical barriers to trade
TDS	total dissolved solids
TRIMS	trade-related investment measures
TRIPs	trade-related intellectual property rights
TSS	total suspended solids
TTSID	Technology Transfer for Sustainable Industrial Development (SDPI)

UNCED	United Nations Conference on Environment and Development
UNCITRAL	United Nations Commission on International Trade Law
UNCTAD	United Nations Conference on Trade and Development
UNCTC	United Nations Centre on Transnational Corporations
UNDP	United Nations Development Programme
UNESCO	United Nations Educational, Scientific and Cultural Organization
UNHCR	United Nations High Commissioner for Refugees
UNICEF	United Nations Children's Fund
USAID	US Agency for International Development
USDA	United States Department of Agriculture
WCED	World Commission on Environment and Development
WTO	World Trade Organization

Introduction

Shahrukh Rafi Khan

The trade and environment debate was partly responsible for the dramatic events that brought different groups out into the streets to disrupt the proceedings of the WTO ministerial meetings in Seattle in November 1999, a phenomenon that has since been repeated at many international meetings. Labour in the North, which is concerned about the loss of jobs to more competitive Southern manufacturing, found common cause with environmentalists who are concerned about Southern imports that result in deforestation, climate change, loss in biodiversity, species loss and other forms of environmental degradation. Southern delegates dragged their feet in protest over the lack of Northern implementation of the GATT Uruguay Round agreements and the exclusionary ministerial negotiation procedures. It was an odd but effective combination that contributed to scuttling the November 1999 WTO talks. Of course, the most important reason for the deadlock was probably the lack of intra-North agreement on issues such as agricultural subsidies.

While diverse Northern and Southern interests coincided in blocking the talks, there are important differences in perceptions on just who the villains are. For Northern labour, it is the multinational corporations that put profit before people and are the conveyors of cheap, labour-displacing and de-industrializing Southern goods. For the environmentalists, it is careless Southern governments as much as greedy multinationals. For Southern governments, it is the powerful Northern governments who represent the interests of their powerful corporations and taxpayers and who are completely unwilling to honour their commitments.

This book examines the conflicts, synergies and policy choices pertaining to the trade and environment interface. Southern governments have dug in their heels in viewing environmental action as too costly to worry about at this stage and think that, in any case, it is part of the Northern agenda of trade protectionism being rammed down their throats.

One of the recurring themes in this book is that attending to the needs of the environment is a multi-win situation from a Southern perspective. First, the costs of environmental inaction are very high in the form of deaths, job loss, working days lost, productivity declines and health care costs. Second, the poor are the most vulnerable in this regard, since the rich have the ability to insulate themselves against dirty air and water and their consequences. Third, environmental action at the micro-level can lead to cost declines resulting from recycling, waste reduction, energy efficiency and other savings resulting from conforming to modern environmental management systems. Fourth, conforming to such standards can lead to macro-gains of avoiding water and resource degradation and also lead to aesthetic gains. Fifth, conforming to such standards can also result in retaining or winning export markets. Finally, the research results reported in this volume show that the costs of mitigation may actually be quite modest and certainly not as high as commonly believed. In view of the above, the South has good reason to engage with the North on the trade and environment issue. Good diplomacy means extracting a price for doing something that is in the Southern interest anyway.

Past trade negotiations and particularly the ministerials have quite clearly demonstrated that the North and South do not represent homogeneous blocks of common interests. In fact, the USA and the EU have differed and continue to do so on various issues such as agricultural subsidies and genetically modified organisms (GMOs). In fact, as indicated earlier, this probably had more to do with the impasse at Seattle than Southern recalcitrance or the labour and environmental lobby and street protests. Similarly, the South represents countries at different levels of development and different interests. Thus it is clearly a simplification to refer to the North and South as representing blocks of countries with unified interests. Even so, we think that, particularly regarding the trade environment debate, a bifurcation continues to be relevant.

This book contains four parts. The first provides an overview of the trade and environment interface, identifying linkages and Southern concerns. The second grounds this overview within Pakistan's context as a case study. Cotton and cotton products and leather exports are selected as two key export commodities. Government and market responses are explored as alternative governance mechanisms at the various stages of the cotton commodity chain. The environmental impact of export enhancement and the cost and benefits of mitigation are identified for both cotton and leather industries. Finally, the move towards cleaner industrial production is described in some detail.

Part III has more to do with investment than with trade. However, foreign investment represents a movement of capital as opposed to a movement of goods. It has important implications for the trade environment interface. On one level, it could be a substitute for imports from the North, if foreign companies produce for the local market. Insofar as global trade has global environmental repercussions via the intensive use of fossil fuels, this is a benefit. However, the South generally pushes for foreign investment as a vehicle for increased exports and in order to address their balance of payment constraints. Two trade- and environment-related investment issues have emerged, and the South is likely to have to take positions on both.

One is the clean development mechanism (CDM) agreed to in Kyoto, which deals with investment in sustainable development in the South by the North and credits so acquired in ameliorating climate change could be traded.[1] This represents an innovative mechanism via which trade can enable the attainment of sustainable development in the South. However, this mechanism is still on the drawing-board and the numerous complexities, including the reluctance of the USA to ratify the treaty, need to be resolved before implementation.

The second is the likely emergence of a new multilateral investment agreement (MAI) that was earlier aborted due to intra-North rivalry, as was the case for climate change negotiations at The Hague. The CDM represents a particular type of environmental investment with a potential impact on the environment and sustainable development. The multilateral investment agreement looks at this issue of foreign investment and its possible impact on the environment and on sustainable development more broadly.

Part IV returns to the primary theme of the trade and environment interface. Practical suggestions are presented for the South with regard to participating more effectively in the trade and environment debate and to the positions the South could take in future rounds of trade negotiations so that the environment is not a stick they are beaten with but a valuable bargaining chip. The book ends with a summary describing the main findings or messages put across by the authors.

Thanks are due to the International Institute for Sustainable Development (IISD) for initiating and generating funding from the International Development Research Centre (IDRC), Canada, for the Southern Capacity Building Trade and Environment project that led to this book. Aaron Cosbey, in particular, and Mark Halle, Konrad von Moltke and David Runnals of the IISD deserve thanks for supporting Phase I of this project.[2] Thanks are also due to all the participants of the project workshop held in April 1999 in Islamabad, Pakistan, and particularly to

Sajid Kazmi for doing a wonderful job in coordinating it. Thanks are due in particular to Fatima Khan and also Irshad Tabassum of the Sustainable Development Policy Institute (SDPI) for patient word-processing assistance. Finally, and most important, it has been a pleasure working with Robert Molteno, editor at Zed Books, because of his sharp insights and numerous valuable suggestions.

Notes

1. Countries could get 'credits' for instituting policies, such as forestation, that absorb greenhouse gases and hence reduce global warming. These credits, allocated by a recognized international authority, could subsequently be exchanged. Thus countries not directly taking action against climate change could buy credits to meet their targets.

2. Phase II was initiated in May 2001.

PART I
. . . .

Overview: The Trade and
Environment Interface

CHAPTER I

The Trade, Investment and Environment Interface

Aaron Cosbey[1]

§ This brief chapter is an overview of the relationship between trade and the environment. It focuses first on the positive and negative effects that trade and economic openness can have on environment and development. It then looks at the positive and negative environmental effects that a trading system can have on trade and development.

The chapter will not examine the clash between the legal regimes set up for the protection of the environment, and those set up to protect the integrity of the world trading system. Such analyses must be preceded by an understanding of the complex web of interactions that bind the areas of policy that such regimes address.

Trade's Effects on Environment and Development

Positive effects The positive effects of trade on both environment and development stem from an increase in national wealth. However, it should be noted that if trade increases inequity by creating wealth that is mostly concentrated in the hands of the wealthy, then it is working against important development objectives. The three ways in which trade is thought to increase wealth are as follows:

ALLOCATIVE EFFICIENCY Trade allows countries to specialize in efficiently producing those items in which they have a comparative advantage. This allows more goods and services to be produced, with a given endowment of resources, by nations that engage in trade. The other side of this coin is that trade restrictions or distortions tend to decrease allocative efficiency. For example, while the USA has domestic industries in sugar or rice, these commodities can be grown much more cheaply elsewhere. The resources used to produce these goods could be more productively used in other sectors, and the resulting products could be used to purchase sugar and rice from abroad. Nevertheless, quotas and perverse subsidies heavily protect both industries.

EFFICIENCY FROM COMPETITION Trade can also generate wealth by exposing domestic firms to foreign competition, forcing them to become more efficient. Firms that are sheltered from competition tend to tie up economic resources in their lethargic attitude towards increasing efficiency. In some cases, better provision of goods can directly serve development objectives, as in the case of telecommunications and other such infrastructure provision. Again, these efficiency benefits are missed where there are trade restrictions or distortions. An important caveat to this effect arises if a particular sector is dominated by international monopolies. Even efficient domestic producers may suffer if exposed to competition from such firms. Thus the presently unfulfilled need for multilateral agreement on competition policy needs to be addressed urgently.

IMPORTED EFFICIENCY Openness to foreign investment or imported technology can bring more efficient methods of process and production. Standardized multinational firms may bring the same level of stringency to all of their locations worldwide. Others will use outdated, less efficient technology in countries where health, safety and environmental protection are more lax.

Where these three mechanisms for creating wealth are at work, there are two ways in which they might support development and environmental objectives. Efficiency may reduce the need for inputs and the production of waste needed to produce any level of output. One of the most frequently cited arguments, for example, is for the reform of the European support system for agriculture. Reform would reduce the over-use of pesticides and fertilizers, and reduce the problems of treating feedlot waste from over-intensive levels of production. Similarly, exposure to competition and import of new technology and know-how may have substantial environmental benefits if the industry is a user of natural resources or a major polluter.

A second type of environmental benefit from increased efficiency derives from the effects of increased wealth. Wealthier people tend to demand better environmental quality – supporting stricter laws and enforcement on environmental concerns, and purchasing costly 'green' goods, which entail less environmental damage. Poor people, who may depend on the environment more directly than the rich, simply lack the means to express the demand. Furthermore, where trade alleviates extreme poverty, it may save people from a vicious cycle whereby they are forced to degrade their own natural environment to survive and, in the process, become increasingly impoverished.

To summarize: trade may achieve development objectives by increasing national wealth through fostering three different types of efficiency. Efficiency may also be good for the environment, since fewer inputs from the environment means less waste in the process of production. Also, wealthier people may demand higher environmental quality and enable the poor to escape a vicious cycle of environmental destruction and impoverishment.

Negative effects There are five ways in which trade and economic openness might work against development and environmental protection objectives:

1. scale effects;
2. income effects;
3. competitive effects;
4. direct effects; and
5. timing and transition effects.

SCALE EFFECTS The scale effect is an example of trade acting as a magnifier of existing problems such as inadequate environmental regulation. The scale effects arise when the trade-induced increases in allocative efficiency increase consumption of goods and services. As the scale of the economy increases, so do attendant environmental problems, natural resource use and the production of waste. This effect occurs only if the existing environmental regulations are inadequate, so that increased production translates into excess environmental degradation. However, no country in the world can claim to price its environmental resources so as to reflect the full costs of their degradation or depletion. The negative impact may be offset by efficiency of resource use from competition or imported efficiency.

It is also sometimes argued that the ability of wealthier citizens to demand higher environmental quality also offsets the scale effect. The argument is the comforting foundation for calls to 'grow now, green later'. Such a strategy has some appeal, but it ignores the fact that preventing environmental damage is far cheaper than cleaning up after the fact. Also, many environmental resources are irreplaceable. This type of damage is exemplified in species loss and the severe non-reversible degradation of fragile ecosystems and renewable resources such as forests and fisheries.

INCOME EFFECTS The income effect is the other side of the beneficial effect noted earlier, and discussed above, whereby increased wealth creates demand for higher environmental quality. However, the richer countries of the world also have far higher percapita emissions of all types of greenhouse gases and toxins than do developing countries. In a nutshell, with enough wealth comes the opportunity to consume wastefully.

COMPETITIVE EFFECTS The competitive effects arise from the advantage inherent in lower environmental standards. There are two related types: pollution haven effects and regulatory chill effects. Pollution havens are countries to which polluting industries are drawn in an effort to escape the costs of environmental regulation in countries implementing higher standards. If, on the one hand, the lower standards accurately reflect differing environmental conditions, then they are a legitimate component of competitive advantage. Even though the result is increased environmental damage, it is outweighed by the resulting development benefits of increased employment. If, on the other hand, they are set lower than they should be according to environmental conditions and the desires of the citizens, then the resulting resource degradation or increase in pollution outweigh the development advantages. There is, of course, no easy answer as to whether a given level of environmental standards is at the 'right' level for a given country.

It has generally been difficult to find much evidence of firm relocation in response to lower environmental regulations. Environmental costs, infrastructure, access to inputs, wage costs, labour productivity, political risk and the broader structure of government regulations must all be taken into account before relocation. The average environmental costs in surveyed firms runs around 2 to 3 per cent of total costs (see Chapter 2), but in certain sectors (such as aluminium smelting or cement manufacturing) it can run much higher.

In the regulatory chill effect, that threat may be explicit or simply anticipated by regulators. An example of the former would be a firm claiming that high environmental standards or strict enforcement are driving it out of business and hence pleading for special treatment. An example of the latter would be government regulators balking at strengthening their environmental laws for fear of driving away existing business, or losing potential business investment. In either case, the environment may suffer if the result is inappropriately low standards or enforcement.

DIRECT EFFECTS Another manner in which trade may be detrimental to environment and development is through direct effects. This is where the trade is in and of itself environmentally damaging or contrary to development objectives. Examples include traded goods that are hazardous waste, endangered species, illicit drugs and prohibited goods.

TRANSITION EFFECTS There is another set of possible negative effects of economic openness related to the timing of liberalization and the transitional effects that follow. These effects result from openness to flows of goods, services, direct investment, portfolio investment and currency speculation. Small developing economies, in particular, may be hamstrung by geographical, sectoral or institutional inflexibilities that cause liberalization to produce painful and protracted periods of transition. Experience has shown that economic openness must be properly staged and accompanied by deliberate domestic policies to facilitate restructuring. In the area of investment specifically, liberalizing while domestic capital markets are weak or immature may leave a country too much at the mercy of unpredictable international capital market trends.

To summarize, by increasing wealth and economic activity, trade may increase environmental damage if existing environmental laws are weak, or if a wealthier population begins to consume wastefully. Trade between high- and low-standard countries may create competitive frictions. Manufacturing relocation from high-standard countries to escape environmental compliance costs is likely. Thus there is pressure on regulators in those countries to weaken environmental regulations to retain or attract business investment. Certain types of trade may have direct negative effects on environment and development, such as trade in hazardous wastes, endangered species, illicit drugs and domestically prohibited goods. Finally, the proper timing or staging of liberalizing trade and investment regimes is crucial in order to avoid a host of negative development effects.

The Effects of Environmental Concerns on Trade and Development

Positive effects There are two ways in which environmental concerns can be expressed through the trading system with beneficial effects on environment and development. There may be an environmental benefit, if exporting firms respond to buyers' or regulators' demands for a greener product. There may be development benefits, if marketing of greener products results in greater or maintained market share.

Exporters who green their operations may do so in response to buyer demand. The pressure to green will usually come from buyers who use or distribute the goods. For example, in the same way that ISO 9000 quality standards have become a prerequisite for international business in many sectors, many buyers are now demanding that suppliers be certified as complying with the ISO 14001 Environmental Management System. Concerned about the image of their final goods among environmentally concerned consumers, these buyers are transmitting demands for greener production down the supply chain. The environmental benefits will accrue in the country of production, as firms use fewer natural resource inputs and produce less pollution. Environmental benefits will accrue in the country of import, since units that are energy-efficient may use fewer resources. Global benefits include reduced emissions of ozone-depleting substances.

Alternatively, exporters may respond to mandatory technical regulations in their target markets that specify product characteristics such as recyclability, recycled content, packaging and labelling requirements, maximum residues of toxic chemicals, and so on. The European ban on the import of textiles and clothing treated with azo dyes is one example of this type of technical regulation. From the perspective of the exporting firm, there is little difference between complying with voluntary standards demanded by buyers and with technical regulations demanded by governments; in either case, the firm must either adapt or lose market share. New technical regulations may cause problems for firms that are unable to change due to a lack of resources, lack of technical or administrative capacity, or where the regulations do not allow sufficient time for transition. The environmental benefits of greening in response to technical regulations will almost always accrue in the importing country, since such regulations usually aim to reduce the environmental impact of the good's use and disposal. Again, some regulations may aim for global benefits, such as regulations on automobile fuel efficiency that are aimed at reducing greenhouse gas emissions.

Still other exporters may, rather than respond to demands from regulators or existing buyers, go beyond regulation to improve unilaterally their environmental performance in an effort to create new market niches and win new customers. A number of coffee producers in Mexico, for example, have collaborated on marketing organically grown coffee, which can be sold at premium prices. In some cases, firms pursuing this strategy may seek to certify the virtues of their products or processes by an eco-label, either self-declared or, more usually, granted in the importing country. Or they may seek to certify themselves as adhering to a high international production standard such as the ISO 14001 Environmental Management System. For many firms, following this strategy is a way to distinguish themselves from other suppliers in an increasingly competitive global market.

The environmental benefits of such strategic greening are obvious. There may also be development benefits in terms of increased employment and income, if firms following such strategies win new market share or avoid losing existing customers. The same sorts of benefits may also accrue when firms adapt to environmental demands from regulators or buyers, if, by doing so, they increase or maintain market share. A cleaner domestic environment will also yield development benefits in terms of improved quality of life in the exporting country.

To summarize, environmental concerns expressed through the trading system can have beneficial environmental effects, including reduced resource inputs and reduced pollution. Firms may green production or products in response to buyer demand, technical regulation in export markets, or as part of a strategy to create green market niches. There may also be development benefits if firms thereby increase their market share, or indirectly through environmental improvements.

Negative effects   There are two ways in which environmental concerns can be expressed through the trading system with negative effects on environment and development: green protectionism and eco-imperialism.

Green protectionism is the deliberate use of an environmental disguise for regulations that are in fact aimed at protecting domestic industry. The case is rarely so clear-cut as this description suggests, most such law originates in a real desire for environmental protection, but industrial lobbies then influence the crafting of the law to their own benefit. Environmental lobbyists, happy to have their objectives addressed, do not object.

Japan's Ministry of Transport has proposed laws on fuel efficiency to achieve its targets for reducing greenhouse gas emissions. Because they are based on the weight of vehicles, the planned rules would affect imports of medium and luxury range cars, a European specialty. By contrast, Japanese cars, even those with higher fuel consumption rates, would escape lightly. This type of sector boosting is the predictable result of forcing the ministry responsible for the sector to regulate its own environmental performance.

To the extent that it is effective in protecting domestic industries, green protectionism denies the wealth-creating benefits of trade to potential exporters. There is a South–North aspect to this issue, in that the typical green protectionist is an industrialized country with high pressure for better environmental regulations. If the affected exporters are developing countries, which arguably have a more critical need to create wealth, the effect is all the more odious.

Eco-imperialism is a term coined by developing countries to describe developed countries dictating to them about how they should behave with respect to their own environments. On the one hand, the types of technical regulations discussed above are not eco-imperialist, since they specify the nature of the product itself – it must be packaged a certain way, or must be recyclable – rather than how the product is produced. That is, they are based on product standards, as opposed to process and production method (PPM) standards. As such, they are aimed at environmental problems in the use and disposal of the product – problems that manifest in the importing country. PPM standards, on the other hand, may constitute eco-imperialism, since they are aimed at changing environmental practices in the country of manufacture (though, as the discussion below points out, the effects might be international).

It should be noted that voluntary standards such as eco-labels and green demands from buyers are not eco-imperialist. Unlike technical regulations, they do not stop the flow of non-complying goods at the border, but rather they merely limit the number of willing buyers.

Neither can we fix the label of eco-imperialism to measures aimed at PPMs that cause global environmental damage. There is nothing imperial about trying to change the PPMs of a neighbouring country whose pollution comes directly to you, whether in the air or through a shared watercourse, although negotiated agreement is obviously preferable to trade measures. The controversial issues here are obvious: how great do the international effects have to be to justify PPM-based restrictions? And to what extent should countries be able to aim PPM-based restrictions at truly global problems, such as ozone depletion, rather than those that affect them uniquely?

Truly eco-imperialist measures may have three types of effect on the environment and development. First, like green protectionism, to the extent that they deny export opportunities, they deny the benefits of the wealth-creating effect of trade. Again, if the victims are developing countries, this is doubly unfortunate. Like technical regulations, they may cause problems for firms that are unable to change due to lack of resources, lack of technical or administrative capacity, or where the measures do not allow sufficient time for smooth transition.

Second, they may in fact achieve some improvement in environmental conditions. This may either come through compliance with the measures, or through negotiation of international environmental agreements designed to remove the need for, or the threat of, the measures.

The agreement on conservation of turtles in shrimp-fishing, signed by the USA and other countries of the Caribbean and Eastern Pacific, had a strong grounding in the fear of US unilateral trade measures should the negotiations fail.

Third, by creating resentment in the countries to which the measures apply, eco-imperialist measures may sour the prospects for cooperation on the ongoing international agenda for sustainable development. Mexican recalcitrance on environmental issues in the World Trade Organization is strongly linked to its resentment of the way the NAFTA and its environmental side agreement have been used to promote the environmental priorities of the USA and Canada in North America. More broadly, eco-imperialism is a rejection of the principles laid down for sustainable development cooperation at the Rio Summit in 1992 – the basis for future work on climate change, biodiversity protection and other important areas of endeavour.

It is worth noting that few pure eco-imperialist measures exist. For one thing, most environmental damage in the production process has at least some international dimension. For another, unilateral extra-jurisdictional measures face tough going in the dispute settlement procedures of the World Trade Organization, which has traditionally ruled them illegal.

To summarize, environmental concerns in the trading system may have negative effects on environment and development if they result in green protectionism or eco-imperialism. Both types of measures can deny developing countries the wealth-creating benefits of trade. Eco-imperialism, moreover, can sour the prospects for North–South co-operation on a number of important sustainable development issues.

Conclusions

If this chapter has one objective, it is to dispel the idea that the trade-sustainable development relationship can be easily described as either negative or positive. It is an immensely complex interaction that varies from country to country, sector to sector, and firm to firm. There are both threats and opportunities in this relationship for countries and firms pursuing economic development and environmental protection.

As in any such situation, then, the imperative is to exploit the opportunities and to reduce the threats. A first prerequisite to doing this is to understand the relationship fully – what are the environmental and development linkages to trade in key sectors, and what are the policy options available? For this reason, continued research undertaken on these issues is of critical importance.

Three trends make it imperative to take strategic action in the area of trade and the environment. First, the world is becoming a more global marketplace, meaning greater competitive pressures. Developing countries will surely face these pressures as the quotas are lifted under the dismantling of the Multi-Fibre Arrangement and traditional competitors gear up their production (see Chapters 3 and 4). Globalization also means that matters once considered purely domestic, such as environmental policy, investment policy and competition policy, are increasingly becoming multilateral concerns.

The second trend is increasing and persistent concern for the environment. This trend manifests itself not only in developed countries, where it is driving the demand for cleaner, greener goods, but also in developing countries, where citizens are becoming concerned at the high environmental price of economic development without environmental safeguards, and are demanding change.

The third trend is the increasing recognition that the trading system needs to address environment and development concerns. WTO dispute panels have been called on repeatedly to try to separate green protectionism from legitimate environmental protection, and many are arguing that there needs to be an explicitly stated process for doing so. The lack of such rules means that dispute panels must reinvent the interpretation of the existing rules on a case-by-case basis. Whether or not this need is addressed in possible new negotiations in the WTO, it is a certainty that environment as an issue will somehow be addressed, and all WTO members need to know in advance where their national interest lie on the issues involved.

Development in the WTO is not a new issue, but the way in which it has been approached here has at times been new. Special and differential treatment for developing countries, the traditional context for development issues in the WTO, is clearly important. But it is also important to focus on how to exploit opportunities for developing countries to profit from the growing markets for greener goods and services.

To repeat a key message in closing, these trends reflect both threats and opportunities for countries and for individual firms. The challenge is to know them better and thereby to help ensure more sustainable development.

Note

1. Interim director, Trade and Sustainable Development Program International Institute for Sustainable Development, Winnipeg, Canada.

. . . .

CHAPTER 2

Trade Liberalization and the Environment: Northern and Southern Perspectives

Shahrukh Rafi Khan[1]

§ The poor Southern countries are currently in a double bind. On the one hand, they find that the rich countries are being very slow in implementing the Uruguay Round trade agreements in liberalizing imports, particularly in sectors of interest to them, such as textiles and agriculture. On the other hand, the world trade scenario is changing, independently of the sway of the WTO, as governments and businesses respond to consumer preferences for ecologically friendly production and consumption and set and impose environmental standards. Thus even the goods currently being exported are increasingly being expected to meet stringent environmental standards.

Poor countries now feel that when it suited the North, they preached consumer sovereignty and confronted them with the 'let the market decide' rhetoric. Now that several countries in the South have acquired comparative advantage in manufactured goods, the North is hiding behind environmental barriers to protect their industries, and setting aside the market ideology they preached.

The issue is not quite as simple as it seems. If standards are responding to consumer preferences in the North, then the market ideology still prevails, and Northern consumers in effect choose to consume goods that are produced by cleaner methods rather than those that are cheaper. However, Southern countries may need to be wary of the protectionist use of environmental standards by rich country governments (see Chapter 1) rather than those dictated by the market. In such cases, they should lobby via the WTO to ensure that the old-time market rule of consumers' sovereignty prevails, particularly now that this benefits the poor countries.

Principle 21 of the *Rio Declaration on Environment and Development* (UNCED 1992: 10)] suggests that international cooperation to create a supportive and open economic system would lead to both economic growth and sustainable development. The document also suggests that trade and environment goals can be mutually supportive (ibid.: 19). Fair market access and prices (without subsidy distortions) for poor countries can generate resources and also ensure the efficient allocation of resources (ibid.: 20). First, these resources can be utilized for sustainable development.[2] Second, the mechanisms of trade itself can enhance sustainable development via cleaner process and production methods (PPMs), with the impetus for this coming from discerning consumers, shareholders and responsive governments.

On 15 April 1994, at Marrakesh, the contracting parties to the GATT put their signatures on the agreement to set up a World Trade Organization (WTO), thus concluding the prolonged Uruguay Round. The first task before the General Council of the WTO, after being set up at the start of 1995, was to constitute a Committee on Trade and Environment (CTE). This reflected the priority attached to bringing environment in the purview of trade discussions. The terms of reference of the CTE were as follows: (a) the identification of the linkages between trade and environmental measures in order to promote sustainable development, and (b) appropriate recommendations on whether any modifications of the multilateral trading system are required. Within these terms of reference, and to promote the UNCED objective of making international trade and environmental policies mutually supportive, an extensive work programme in ten areas was decided upon and initiated in a specially set up sub-committee of the Preparatory Committee of the WTO.[3] The centre-stage of the international debate on development in the remaining part of the 1990s and the early part of the next century is likely to be occupied by the issues of trade, environment and sustainable development.

Traditional trade theory, based on the concept of 'comparative advantage', claims that trade brings mutual benefits to all parties engaged in exchange. However, the theory of comparative advantage assumes that all external costs are internalized, when typically they are not. The terms of trade of a country thus do not reflect the social costs involved in the production and consumption of goods and services to be traded.

The trade and environment literature deals with a number of other issues and hypotheses that are not a part of traditional trade theory. Many of these are related to concerns in the North or the South about fair trade. First, that trade liberalization could result in strategic movement on the part of Northern multinational corporations to Southern countries with more lax environmental regulations and hence result in a loss in Northern jobs. Second, that the North could use trade liberalization to dump its dirty technology and other domestically prohibited goods (DPGs) on the South. Third, that the structural adjustment induced export promotion could result in the South exporting its environmental capital in the form of high domestic pollution and resource degradation. Fourth, that the multilateral environmental agreements (MEAs) are increasingly affecting the world trading environment and these MEAs could block Southern exports. Fifth, that the North has a greater resource and technological ability to meet the standards it sets and that this will mean blocking access to Southern exports and enhancing its market share. Sixth, that the cost of mitigating such pollution in the South is very high. The literature on these issues is reviewed in greater detail in the next section.

Northern and Southern Positions

Loss of Northern jobs Companies in the North may fear that, with the dismantling of trade barriers, developing countries may have a competitive edge due to their less stringent or more lax enforcement of environmental regulations. This might lead to a relocation of factories to developing countries to take advantage of lax environmental regulations and/or enforcement. Repetto (1993), Dean (1991) and Tobey (1990) refute this hypothesis. They argue that relocating a plant entails complex and lengthy processes, which include selling an existing plant, severing its workforce, relocation of key personnel, choosing a new site, building a new factory, and recruiting and training new staff. All these processes are not feasible just to take advantage of savings on pollution control cost that total less than 2 per cent of total sales. The *World Development Report* (World Bank 1992: 67) also states that environmental costs are a minor share of output value – averaging only 0.5 per cent for all US industries in 1988 and 3 per cent for the most polluting industry.

Mani and Wheeler (1997) find, using cross-country analysis, that the evidence seems consistent with the pollution haven pattern of investment. However, upon closer examination, they suggest that there are several other reasons explaining 'dirty production' in the South that have little to do with the 'pollution haven' story.

Imports of 'dirty industry' into the South Developing countries feel threatened that, with trade liberalization (that is, reduced tariffs on imported capital and intermediate goods), there may be an influx of dirty technology coming into their countries. While evidence on this is limited, there was an instance in Pakistan in which a second-hand Danish mercury chlor-alkali plant was being imported in 1994.[4] Greenpeace International, with the support of local environmental organizations, frustrated this attempt. Similarly, the newspapers reported on the proposed dumping of toxic wastes off the coast of Pakistan's Balochistan province. Thus the world environmental community needs to be alert to the disposal of various domestically prohibited goods (DPGs) including 'dirty machinery', toxic wastes, insecticides, fertilizers, chemicals and pharmaceuticals.[5]

Exporting the environment Critics of the free trade ideology claim that increased exports, particularly in the aftermath of liberalization, will be at the cost of natural resource depletion and degradation and increased industrial pollution. Thus the World Commission on Environment and Development (1987) pointed out, in what is referred to as the Brundtland Report, that, during the 1980s, the South's commodity trade was based on the over-harvesting of nature in order to service its debt. The problem is especially acute in that the South lacks the resources and technological prowess to combat environmental degradation.

Proponents of liberalization argue that, quite to the contrary, enhanced exports are likely to benefit the environment in the long run. Birdsall and Wheeler (1992) point out that competition will induce the drive towards the latest manufacturing technologies and that, since these are likely to be procured from the North, they are likely to be much cleaner. Further, Northern importers may require cleaner processes to ensure greener products.[6] They present evidence from their own cross-country analysis showing greater openness to be associated with less pollution-intensive industrialization.[7] Eliste and Fredriksson (1998) find that, for the agricultural sector, trade liberalization does not induce a 'race to the bottom'. Their findings suggest a positive relationship between stringency of environmental regulations and trade openness. Their findings also suggest that there is a positive association of the degree of stringency in regulations among trading partners.

Cross-country evidence can at best be viewed as suggestive. Thus more evidence on this issue, based on industry case studies, is awaited. Dean (1998) developed and estimated a simultaneous equation model for Chinese provincial data to show that the direct effect of liberalization, via the terms of trade, is negative but the indirect effect via income growth is positive. Again, the income growth effect could equally be neutral and essentially depends on the political economy of resource allocation in a particular setting. Strutt and Anderson (1998) develop a methodology to study the impact of trade liberalization and environmental depletion and apply this to Indonesia. They find that trade policy reform expected in the next two decades would, in many cases, given the current state of environmental regulation, improve the environment and reduce resource depletion with regard to air and water. Citing other research that they have done, the same is claimed for land degradation via soil erosion and associated off-site damage. In the worst-case scenarios, trade liberalization is expected to add only slightly to environmental degradation.

MEAs as a tool of protection In recent years, trade policy has been considered as an instrument to enforce environmental compliance in the form of inclusion of trade provisions in multilateral environmental agreements (MEAs).[8] These may include unilateral use of trade measures to enforce environmental compliance on the part of trading partners. The sanctions, if applied, would mean that trade with non-parties to the agreement would in principle be prohibited. So far the WTO has not endorsed the use of such sanctions. Nevertheless, these MEAs are an important feature in the trade–environment interface.

The provisions of the Montreal Protocol required signatories to ban imports of CFCs (chlorofluorocarbons) and products containing CFCs from non-signatory countries. Precedence now exists regarding the unilateral use of trade measures to enforce environmental compliance – the US ban on shrimp imports to encourage the use of turtle-excluder devices to protect sea turtles is a case in point.[9] The Convention on International Trade in Endangered Species (CITES) has agreed to a ban on ivory. Other countries have import bans on whales, fur seals, polar bears and some specific migratory birds and species. The Basle Convention bans some types of trade in hazardous and toxic wastes.

The North has a comparative advantage in meeting environmental standards The emphasis by the Northern environmental community on uniformity of production and process methods and environmental effects of production processes is interpreted by the South as an effort to restrict its access to Northern markets. The argument is that Southern countries do not have the capacity of Northern countries to cope with detailed regulation, and also that the regulations are tailored to Northern concerns and may thus be inappropriate. Thus any benefits of liberalization and environmental conservation, in the presence of harmonized standards, will be skewed in favour of the North.[10] Brazil raised this issue originally in 1993 over European Union regulations for tissue paper production. Brazilian pulp manufacturers complained that the regulations on consumption of renewable and non-renewable resources, waste generation and sulphur emissions would disadvantage foreign producers who could not meet these standards.[11] Kaushik (1999) reviews various cases for India and finds that the high standards are so rigorous in some cases that they could be viewed as protectionist. Further, sometimes they may be motivated by Northern producers wanting to market alternatives. Not only in the case of compliance are costs prohibitive, but legitimate questions regarding environmental justification can be raised and there is often an irrelevance of Northern standards. Finally, there is a lack of concession to local Southern conditions, such as the supply of sustainably produced wood in the supplier's market.

High mitigation costs in the South   Many Southern countries exporting to the OECD countries have had to confront standards, particularly in the leather and textile industries, and this is viewed as an unfair protectionist cost being imposed on them by Northern governments.[12] Our take on this issue differs. Southern countries such as Pakistan must distinguish between restrictions imposed by Northern governments and those imposed by Northern businesses. If Northern governments impose import restrictions because Southern countries are not doing enough about child labour or cleaning up production technologies, this constitutes a non-tariff barrier. However, this is not the big danger that faces Southern exporters. Increasingly, businesses in the North are being required by their boards/shareholders to do business with firms that meet certain 'voluntary' environmental and quality standards. In some ways, a cleaner environment is viewed as a luxury good and the more prosperous Northern consumers are viewed as requiring it.[13] This is thus a market-dictated standard and not as such a non-tariff barrier imposed by Northern governments. This is a very important distinction. The only option Southern exporters have is to conform or lose markets.

Even if Northern governments impose standards and they provide an edge to Northern producers who are more capable of meeting them, it would still be wise for LDCs to conform. Various product-related environmental standards should be seen as an ongoing consumer protection movement in the North. While various process-related standards can legitimately be viewed as impediments to trade, as long as governments impose these (see Chapter 1), it is difficult to argue with consumer sovereignty in the North. Further, based on research subsequently described in this volume (Chapter 4), our view is that cleaning up production processes generates far more social benefits than costs in producer countries, and wins markets as well.

Notes

1. The author is executive director of the Sustainable Development Policy Institute, Islamabad, Pakistan. Assistance from Haider Ghani and earlier work on this issue at the Sustainable Development Policy Institute are gratefully acknowledged.

2. There is of course no guarantee that this will happen.

3. Initially, seven items were on the list and three (services and the environment, TRIPs and the environment, and relationships with other institutions and organizations) were later added to make ten. Meecham (1998: 87–90) provides an account of the recent history of trade and the environment and also the trend in thinking within the CTE on various issues (ibid.: 94–109). A good source for the latter is the regular WTO *Trade and Environment Bulletin*. Refer to UNDP/UNCTAD 1998: 24-7 for the mandate of the CTE.

4. Mercury–based production of chlor-alkali will be phased out by Paris Convention countries, of which Denmark is a member, by 2010. Also Jha and Teixeira (1997: 179) note the movement of leather-tanning to the South as the North imposed stringent environment standards.

5. OECD 1994.

6. They cite evidence of German imports of fishmeal and paper products from Chile that required treatment of effluent to ensure reduced bacteriological contamination of products (1992: 160). Another example, cited by Robins and Roberts (1997: 22), is the production adjustment of Indian textile producers to the ban on azo dyes.

7. Wheeler and Martin (1992) present evidence of greater openness leading to cleaner technologies due to competitive pressures in the case of wood pulp production.

8. Cai et al. 1997: 21.

9. Even as the WTO dispute settlement procedures struck down this unilateral US action, which is also in violation of Principle 21 of the Rio Declaration, the USA moved ahead with enforcing an import ban on all but 37 certified countries. The Pakistani media reported on this act of unilateralism in 'US bans shrimp imports from Pakistan, India', *The Nation*, 6 May 1999 and 'Trawler owners asked to comply with rules', *Dawn*, 23 May 1999.

10. Nath 1997.

11. *Ecologist*, Vol. 25, No. 1, 1995.

12. The WWF (1997) points out that India, China and Zimbabwe confronted barriers due to textile dies. Refer to CBI/CREM *Environmental Quick Scans* for identifying bans, standards and existing and intended environmental legislation applicable to EU imports from developing countries.

13. This positive income elasticity for a cleaner environment is the logic underlying the controversial environmental Kuznet's curve. Refer to Grossman and Krueger (1991).

Bibliography

Birdsall, N. and D. Wheeler (1992), 'Trade policy and industrial pollution in Latin America: where are the pollution havens?', in Patrick Low (ed.), *International Trade and the Environment*, World Bank Discussion Papers, No. 159.

CBI/CREM (Centre for the Promotion of Imports from Developing Countries/ Consultancy and Research for Environmental Management (1998), *Environmental Quick Scans: Textiles: A Trade Related Orientation on Environmental and Health Issues Relevant to Exporters to the EU*, Rotterdam.

Cai, W., M. Isolda, G. Guevara and C. Hamilton (1997), 'Pakistan and the Uruguay Round: issues, implementation and impact', Centre for Trade Policy and Law, Norman Paterson School of International Affairs, Carlton University, Occasional Paper in International Trade Law and Policy, No. 44.

Dean, J. M. (1998), 'Testing the impact of trade liberalization on the environment: theory and practice', World Bank Conference on Trade, Global Policy and the Environment', Washington, DC, mimeo.

Dean, J. M. (1991), 'Trade and the environment: a survey of literature', Background paper prepared for the *World Development Report 1992*, World Bank, Washington, DC.

Eliste, P. and Per G. Fredriksson (1998), 'Does open trade result in a race to the bottom? Cross-country evidence', World Bank Conference on Trade, Global Policy and the Environment, Washington, DC, mimeo.

Grossman, G. and A. Krueger (1991), 'Environmental impacts of the North American Free Trade Agreement', Princeton, NJ: Princeton University Press.

Jha, A. and A. P. Teixeira (1997), 'Transfer of environmental sustainable technology', in V. Jha, G. Hewison and M. Underhill (eds), *Trade, Environment and Sustainable Development: A South Asian Perspective*, London: Macmillan.

Kaushik, A. (1999), 'Promoting sustainable trade: the case of environmental requirements in India', paper presented at an IISD/SDPI/IUCN workshop on Trade and Sustainable Development, Islamabad.

Mani, M. and D. Wheeler (1997), 'In search of pollution havens? Dirty industry in the world economy, 1960–1995', World Bank Conference on Trade, Global Policy and the Environment, Washington, DC, mimeo.

Meecham, R. (1998), 'The Indian Ocean, the WTO, trade and sustainable development', in D. Shah Khan (ed.), *Pakistan: To the Future With Hope* Karachi: SDPI/Vanguard.

Nath, K. (1997), 'Trade, environment and sustainable development', in V. Jha, G. Hewison and M. Underhill (eds), *Trade, Environment and Sustainable Development: A South Asian Perspective*, London: Macmillan.

OECD (1994), *The Environmental Effects of Trade*, Paris: OECD.

— (1996), *Reconciling Trade, Environment and Development Policies: The Role of Development Co-operation*, Paris: OECD.

Repetto, R. (1993), 'Trade and sustainable development', Second Asian Development Bank Conference on Development Economics, ADB Manila, November 1993.

Robins, N. and S. Roberts (1997), *Unlocking Trade Opportunities*, New York: International Institute of Environment and Development and UN Department of Policy Co-ordination and Sustainable Development.

Strutt, A. and K. Anderson (1998), 'Will trade liberalization harm the environment: the case of Indonesia', World Bank Conference on Trade, Global Policy and the Environment, Washington, DC, mimeo.

Tobey, J. A. (1990), 'The effects of domestic environmental policies on patterns of world trade: an empirical test', *Kyklos*, Vol. 43, No. 2.

UNCED (1992), *Agenda 21: Programme of Action for Sustainable Development/Rio Declaration on Environment and Development*, New York: United Nations Department of Public Information.

UNDP/UNCTAD (1998), *Trade and Environment: Capacity Building for Sustainable Development*, New York.

Wheeler, D. and P. Martin (1992), in P. Low, *International Trade and the Environment*, World Bank Discussion Papers 159, Washington, DC: World Bank.

World Bank (1992), *World Development Report 1992*, New York: Oxford University Press.

World Commission on Environment and Development (1987), *Our Common Future*, Oxford and New York: Oxford University Press.

PART II

Potential and Actual Responses: Case Studies

. . . .

Alternative Public Regimes for Achieving Environmental Improvement in the Global Cotton Commodity Chain: The Case of Pakistan

Tariq Banuri[1]

§ The purpose of this chapter is to explore prospects and mechanisms for a transition to sustainable development in Pakistan, and more generally in the South, and the effect of international trade on such prospects. The specific case examined here is cotton and cotton products, which together constitute the largest economic sector in Pakistan, with considerable trade exposure at every stage of production. It is a global activity and has long been the subject of North–South trade negotiations, particularly over trade and the environment.

Recent years have seen growing concern in industrialized countries about the environmental impact of cotton production and processing. Conventional production methods (of cotton as well as textiles) are associated with significant and avoidable environmental or health-related costs.[2] Earlier, the motivation for the concerns was the health of farmers and workers, the quality of soil and water, and local biological diversity. The signing of the GATT Uruguay Round added other concerns, in particular, the affect on consumers of carcinogenic dyes and chemicals (especially azo dyes), and production processes resulting from inadequate environmental safeguards and standards that create a competitive disadvantage for 'cleaner' industries in the North.

Analytical blinders hamper these production processes. Some approaches are technology-driven and assume that the only obstacle to sustainable production is the lack of knowledge of technological alternatives. Others assume homogeneous producers, equally capable of entering niche markets, altering their technological and managerial systems, and managing brand names. Far from being neutral, all of these favour the strongest segments of the chain, namely modern producers, operating at a relatively large scale, with preferential access to credit, technology and governmental resources.

The reality is substantially different, albeit quite varied. In Pakistan, the cotton sector exhibits considerable diversity in terms of virtually every characteristic: unit size, formality of industrial structure, nature of competition, and underlying cultural and governance systems. At one extreme are 1.3 million cotton farms (roughly half of these smaller than two hectares) competing in almost classic perfect competition, a vast majority operated as family farms by owners/cultivators with limited literacy or access to technology. The main determinant of technological change in this sector is the governmental system of research and extension – which has become quite ineffective over time. At the other extreme are large-scale textile processors and small-scale garment manufacturers, both subject to the influence of large international corporations, and indeed part of an internationally governed commodity chain. In the middle are large-scale spinning units and small-scale, informal sector weaving units, the latter numbering in the tens of thousands, mostly operating as family enterprises, with virtually no recognized system of industrial governance.

Without an understanding of the underlying system of governance, it is impossible to identify fruitful forms of intervention. To further such understanding, we use the framework of global commodity chains, especially the concept of industrial governance, as articulated by Gereffi (1994) and his colleagues.

The data for the study were collected from a number of official statistical publications, as well as a questionnaire survey and direct interviews with leading experts and practitioners around the country. The chapter makes the following arguments:

- The responsiveness of individual actors in a production chain to changing incentive structures is linked closely to the nature of governance in the chain. In the absence of an effective system of governance (as in cotton production, ginning (separating seeds from cotton) and spinning), transition costs are likely to be high as well as inequitably distributed.
- Given the nature of the commodity chain at the upper end of the market, the costs of transition will fall disproportionately on the manufacturers, while the benefits of changing consumer preferences will accrue to mass retailers who have a comparative advantage in labelling, packaging, advertising, certification and inspection.
- The switch would be impossible without ready access to clean

production technologies. Alternative technologies exist in some areas (such as processing), and are in an experimental stage in other areas (such as green cotton). In the remaining areas (such as integrated pest management (IPM)), while the meta-technology has been well developed, its contextual application and dissemination is not anywhere near the range of feasible options.

• A feasible programme must seek to intervene in existing governance systems, and either strengthen them in order to facilitate the transition, or to transform them through investment and technical assistance to produce an orderly and equitable transition to sustainability.

The global cotton commodity chain: governance and change   The cotton commodity chain can be divided into three broad stages: production, processing and marketing. While conditions in each stage exhibit certain similarities, there are significant differences as well. In particular, while the governance structure of apparel manufacture is controlled by mass retail firms, and that of cotton growing by governmental institutions, albeit increasingly ineffectively, yarn and cloth production do not exhibit a coherent governance arrangement. Yarn production is in the large-scale organized sector, where the manufacturers' association can operate as a lobby to influence government policies. The weaving sector, however, is fairly anarchic in its composition. These differences are critical in evaluating the effectiveness of intervention in production systems.

The textiles and apparel industry is driven by large retailers rather than producers or processors. According to Gereffi (1994) commodity chains have three main dimensions: 1) an input–output structure, 2) territoriality, i.e. spatial dispersion or concentration of enterprises of different sizes and types, and 3) a governance structure, i.e. the authority and power relationships that determine how financial, material and human resources are allocated and flow within a chain. The purpose is to examine the nature of the relationships between various economic agents in order to understand the sources of stability and change.

The difference between producer-driven and buyer-driven commodity chains is primarily in terms of their governance structures. Producer-driven chains are defined as those in which:

trans-national corporations or other large industrial enterprises play the central role in co-ordinating the production system. This is most characteristic of capital-intensive and technology-intensive industries like automobiles, computers, aircraft, and electrical machinery. Buyer driven chains refer to those industries in which large retailers, brand name merchandisers, and trading companies play the pivotal role in setting up decentralised production networks in a variety of exporting countries, typically located in the 'third world'. (Gereffi 1994: 97)

The main job of the core company in buyer-driven chains is to manage production and trade networks. Profits derive not from scale economies and technological advances, as in producer-driven chains, but rather from unique combinations of high-value research, design, sales, marketing and financial services.

This distinction, and in particular the concept of governance, is helpful in examining the commodity chain and exploring the prospects for transformation of production processes. Governance can be described in a number of ways. A good definition is to think of it as an arrangement in which collective decisions can be made in a legitimate manner and with the minimum of conflict. As Gereffi has reminded us, in a market context, governance is derived from market power. Market power, meaning the ability to control or influence markets, is mainly the result of industrial structure. Enterprises in oligopolistic or monopolistic industries have the ability to influence prices and, as a result, generate excess profits. In other words, market power is linked closely with economies of scale, barriers to entry, and nature of industrial concentration.

Market power is self-reinforcing, in the sense that it provides actors with the ability to maintain their preferential position by actions that help create and sustain a favourable market environment. Typically, this has meant the control of one's competitors through collusion or price-fixing (for example, the threat of predatory pricing), the control of the policy environment by 'managing' governmental actors, and the control of consumer choices through advertising and brand name development. All these actions create a bias in favour of large-scale and established enterprises, even when the technology itself is scale-neutral.

The concept of governance provides an additional dimension to such strategic action. This means managing one's suppliers or other intermediaries in such a way as to enhance total sector profits.[3] Such management requires strategies for the distribution of benefits as well as threat of costs. Naturally, this requires the investment of financial and human resources in the management of economic relations with trade partners. It also means the creation of the ability to introduce changes in behaviour, either by the promise of rewards or the threat of sanctions. Rewards and sanctions refer mainly to access to credit, markets, technology and know-how. Given the relationship of scale and profitability with the generation of such surplus resources, it is not surprising that producer-driven chains are characterized by strongly oligopolistic production structures, whereas buyer-driven chains are dominated by monopolistic or oligopolistic retail firms and brand name companies.

However, the idea of market-based governance is relatively new, and is in a sense the 'privatization' of activities that were traditionally the sole domain of government. Even now, wherever market power or an attitude towards governance is not in evidence, the only provider of governance remains the government. As such, to add to Gereffi's categories, one could equally think of a 'government-driven' commodity chain. In this type of a chain, the flow of financial, material and human resources through a chain is influenced strongly by the policies and actions of government agencies. The agricultural system in Pakistan exhibits this kind of structure. The combination of agricultural research, extension, policy, credit and input provision is used to influence the allocation decisions of farmers. As we shall observe in more detail below, this system has deteriorated over time, and is unable to perform its responsibilities effectively. It has also become somewhat captive to the actions of input suppliers.

Finally, one could also think of an 'anarchistic' or 'governance-less' commodity chain, where there is an absence of governing arrangements between producers. Typically, the informal sector in many Southern countries exhibits such behaviour. In Pakistan, the production of cloth is almost entirely in the informal sector.

The special condition in the farm sector, although it is perhaps not central to the segments examined by Gereffi, is that of the system of knowledge underlying production arrangements. Marglin (1990), in his analysis of the organization of work among informal sector weaver communities in eastern India, distinguishes between two approaches to knowledge. He terms these *episteme* and *techne*. The former refers to knowledge that is cerebral, axiomatic and universal, transmittable in transparent and internally egalitarian ways. The latter is tacit or embodied knowledge, contextual and personal, typically transmitted through experience and prolonged exposure, often through hierarchic arrangements. The two embody contrasting processes of innovation, diffusion and legitimacy. *Episteme* derives its legitimacy from impersonal research practices, and *techne* from personal trust.[4]

In the cotton commodity chain, the knowledge base in segments that correspond most closely to perfect competition corresponds most closely to Marglin's *techne*, while that of the large-scale and modern segments is *epistemic*. The result is that intervention in perfectly competitive segments requires an alternative sensibility, an approach to knowledge as personal and based on trust, constantly seeking legitimacy.

To revert to the main theme of this chapter, the analysis that follows will use the concept of industrial governance to motivate the discussion. This concept suggests that the prospects of transformation of industrial methods depend fundamentally on the nature of governance in that sector. To summarize, we could divide the cotton commodity chain into three broad segments:

- *Apparel and textiles* This segment is characterized by buyer-driven commodity chains and strong governance arrangements. Here, the prospects for industrial transformation are relatively bright. However, the main issue is that of equity, support for equitable distribution of costs and benefits from the indigenous processes of transformation. Intervention in small-scale production will require personal and face-to-face interaction. The existing programme of sustainable industrial development offers prospects of linking up with international processes to determine more equitable and efficient outcomes.
- *Yarn and grey cloth* This segment is characterized by relatively weak forms of governance, albeit with significant differences between them. Yarn production is in the large-scale sector, where the manufacturers' association acts as a lobbying group, while weaving is in the informal sector without a coherent collective structure. The prospects of transformation are more muted here, and precipitate actions will involve considerable social and economic inequities as well as avoidable costs. The agenda here is the institution of governance arrangements. However, the environmental costs of these stages of production are smaller than those of the others, and addressing these could be viewed as a longer-term programme.
- *Cotton growing* This segment is still characterized by government-driven initiatives built upon a vast complex of research, extension, input supply and credit. These arrangements have become weak and non-optimal over time. Here the agenda is the restoration of the system of governance by investing in the research and extension network. Alternative networks will have to deal with the obstacle of developing trust and language to communicate with farmers.

Trends in the cotton industry[5] Cotton is the largest revenue-earning non-food crop produced in the world. Its production and processing provide some or all of the cash income of over 250 million people worldwide, including almost 7 per cent of the available labour force in developing countries. These activities are becoming highly concentrated over time: today, 77 per cent of global cotton output and 73 per cent of the cotton hectarage are accounted for by China, the USA, India, Pakistan and the Central Asian Republics. Other features of the sector are:

- Cotton cultivation covers nearly 33 million hectares, equivalent to about 2.5 per cent of all cultivable land, in 82 countries. Southern countries produce 77 per cent of the world's cotton, and constitute 58 per cent of world cotton exports.
- Cotton textiles constitute approximately half the total textile fibre and arguably the largest industry in the world. It has been the leading industrial sector in many Southern economies.
- Cotton trade in the 1990s averaged about 6 million tonnes annually, representing one-third of the crop output, the remaining two-thirds either consumed domestically or exported in processed form.

Because of shifting comparative advantage, there has been a rapid expansion of the textile industry in the South, especially in cotton-growing countries. Textile production has traditionally been the first industrial sector of many developing countries, and has paved the way for broad-scale industrialization and economic expansion. The ten largest cotton-producing countries (of which only one is in the group of industrialized economies) consumed 50 per cent of the global cotton output in 1986, and 77 per cent of a larger volume in 1996 (FAO 1997, cited in IISD/WWF 1997: 52). Similarly, the percentage of cotton traded internationally fell from 38 per cent in 1960–61 to 27 per cent in 1992–93.

Another stylized fact of the market for cotton and cotton products is the existence of special trade barriers against Southern industrial products. Textiles are a labour-intensive industry, and provide a comparative advantage to Southern producers. However, the shift of the industry to the South has been slowed down, and the interests of traditional Northern manufacturers protected under various unilateral, bilateral and multilateral agreements. As early as 1935, a voluntary export restraint on textile exports to the USA was announced (de Vries 1995: 15). In 1974, the Multi-Fibre Arrangement (MFA) was adopted, which tied textile imports to quotas for individual countries. The elaborate systems of quotas set up under the MFA is only now being dismantled through the Agreement on Textiles and Clothing (ATC) negotiated in the GATT process. Under this agreement, products covered by the MFA will be reintegrated into mainstream WTO discipline over a ten-year period beginning in 1995. This is being done in four stages. Sixteen per cent of the 1990 volumes of products covered by the MFA were integrated at the signing of the WTO, another 17 per cent by 1998, 19 per cent by 2001, and the rest, 49 per cent, by 2005 (see also Chapter 4).

Critics have noted that the agreement is heavily biased in favour of importing countries (Low 1995). First, the end loading of the phase-out means that the benefits will not be realized for several years. Second, the agreement gives maximum flexibility to importing countries to select how they will meet their phase-out commitments. Third, even at the end, tariffs on textiles and clothing will be liberalized by only about 22 per cent as compared with 40 per cent overall in the Uruguay Round agreement. The bias is particularly acute for South Asian exporters, who face tighter restrictions on their exports, and although the gap will become narrower, it will remain significant.

Partly as a result of the ATC, and also because of changing market conditions and consumer tastes, the textile market has undergone a significant transformation in recent years. The role and profits of large retail corporations and brand name managers have become far more important. At the same time, the number of players has declined, and industrial concentration in these components of the market has increased. The large corporations have in turn developed closer inspection and other arrangements with their suppliers, who tend to be small scale and decentralized.

In the USA, these include discounters (Wal-Mart, Kmart), mass merchandisers (Sears, Dayton Hudson, Woolworth), department stores (J.C. Penney, May), speciality stores (Melville, The Limited, The Gap, Toys 'R' Us), with sales running into tens of billions of dollars (Gereffi 1994: 106). Similarly, large European retail companies (Bo Weevil in the Netherlands, Otto BV in Germany, Hennes and Mauritz in Sweden and Stockmann in Finland) also have a strong influence on market conditions, although their markets are not as large as those of US retailers.

These retail companies have gradually got into active overseas buying through agents or buyers. Some also provide technical assistance to producers in order to develop speciality products for niche markets. They engage in inspection and certification (generally for their own internal purposes) to ensure that their suppliers meet the labour standards and other legal requirements of the importing country.[6]

The available evidence indicates very high retail margins for the big firms. Gereffi reports 48 per cent average retail mark-ups for US fashion products in 1990, with only one-third of the final value of output going to overseas producers. While similar evidence for mass-produced goods or low-end products is not available, there is an indication that garment manufacturers have been under considerable pressure from retailers to lower prices and speed up delivery schedules. The competitive nature of the supplier segment and the increasingly concentrated nature of the retail segment make this possible. To a certain extent, quotas have limited the pressure under the MFA. Given this, it is not clear whether the gradual elimination of the MFA will benefit Southern producers.

Another important feature is the rise of new concerns, especially those pertaining to environmental and social conditions. As is inevitable in a commodity produced on such a large scale, there are a significant number of negative environmental effects. Cotton plants are susceptible to a large variety of pests and diseases that can cause stunted growth, poor colour, lower yields or even death. Accordingly, there is a long history of pest control practices in cotton production. Traditional methods included a variety of labour-intensive practices (hand-picking of pests, inter-cropping, crop rotation and the burning or removal of cotton residues from the soil). Increasingly, however, over the last 100 years, these methods have been largely forgotten, and have been supplemented or transplanted by reliance on chemical pesticides.[7] Pesticide use in cotton alone is valued at US$2 to 3 billion annually, which is one-quarter of the total insecticide consumption in the world. It has become a significant proportion of production costs, and constitutes close to one-tenth of the annual value of the cotton crop of US$30 billion (Murray 1994).

Besides this, environmental problems are also associated with the prevailing agricultural practices as well as with the use of other chemicals in the growing as well as the processing stage. Water use and common tilling practices effect water quality adversely, and lead to water scarcity, soil erosion, waterlogging and salinity. Chemical fertilizers cause soil and water contamination and affect soil fertility. Some chemical dyes and chlorinates used in the processing stage have carcinogenic effects; others have adverse consequences for human health and water quality if discharged without proper treatment. In other stages of industrial production, although the degree of wastage is fairly low even in traditional production systems, and has been reduced even further in modern methods, problems remain with regard to worker health and safety from suspended particulate matter, especially in smaller plants. These environmental problems are addressed in more detail later in the chapter.

The role of cotton in Pakistan   Pakistan is the fifth largest producer of cotton in the world, the third largest exporter of raw cotton, the fourth largest consumer of cotton, and the largest exporter of cotton yarn. About 1.3 million farmers (out of a total of 5 million) cultivate cotton over 3 million hectares, covering 15 per cent of the cultivable area in the country. Cotton and cotton products contribute about 10 per cent to GDP and 55 per cent to the foreign exchange earnings of the country. Taken as a whole, between 30 and 40 per cent of the cotton ends up as domestic consumption of final products. The remainder is exported as raw cotton, yarn, cloth and garments.

Cotton production supports Pakistan's largest industrial sector, comprising some 400 textile mills, 7 million spindles, 27,000 looms in the mill sector (including 15,000 shuttle-less looms), over 250,000 looms in the non-mill sector, 700 knitwear units, 4,000 garment units (with 200,000 sewing machines), 650 dyeing and finishing units (with finishing capacity of 1,150 million square metres per year), nearly 1,000 ginneries, 300 oil expellers, and 15,000 to 20,000 indigenous, small-scale oil expellers (*kohlus*). It is by any measure Pakistan's most important economic sector. Not surprisingly, government policy has generally been used to maintain a stable and often relatively low domestic price of cotton, especially since 1986–87, through the imposition of export duties, in order to support the domestic industry.

In terms of cotton content, the bulk of the exports take place in the form of cotton yarn, of which about 45 per cent is destined for other countries. During the 1990s, the decline in cotton yields in some years forced net imports of cotton to meet the domestic spinning demand of about 1.4 million tonnes. Indeed, in 1993–94, although 0.3 million tonnes of cotton were imported to meet local demand, 150 mills had to close down because of a shortage of raw materials. Exports of finished textiles have begun to expand since 1990. In recent years, the industry has been jolted by some trade restrictions based on environmental considerations. The most important of these pertains to the ban imposed by European countries on products made with azo dyes, which are feared to have carcinogenic properties.

In fact, the significance of the cotton sector far exceeds its contribution to GDP or exports. The rate of economic growth is correlated quite closely with the fate of the cotton crop. During the 1950s, when cotton yields were stagnant, economic growth was at best lukewarm. In the 1960s, as cotton yields rose by over 50 per cent, GDP growth also reached its highest decadal level of over 7 per cent per annum. The 1970s were turbulent years for the economy, the growth rate declining to 4 per cent per annum; it was also a period of fluctuating but stationary cotton yields. Growth reverted to over 6 per cent in the 1980s, at a time when yields began to increase dramatically, more than doubling in a decade. Finally, since 1991–92, both cotton yields and GDP have become sluggish. In other words, economic activity in the country is associated closely to performance of the cotton crop. A bigger crop means not only a larger volume of exports (both raw and processed products), but also a subsidy to the textiles sector, leading to higher aggregate demand, higher employment, larger fiscal inflows, less pressure on the balance of payments, and thus less exposure to the dictates of international financial organizations.

These trends also exhibit an association with political conditions, albeit not a simple one. The cotton crisis and the resulting economic instability of the 1970s were associated with political instability and growing unpopularity of the government. Likewise, the second cotton crisis of the 1990s and the accompanying economic decline has seen the dismissal of two governments and growing public disenchantment with government in general. The impetus provided by the cotton sector to the economy in the 1980s could be argued to have shielded an otherwise unpopular regime from overt civic unrest. However, the relationship is somewhat more complex than this, and rural unrest became high precisely when agricultural yields were rising dramatically in the 1960s and 1980s.

The commodity chain in Pakistan Briefly, the production stage covers a sequence of activities from sowing to harvesting and ginning.[8] Cotton is produced on large as well as small farms, with significant differences in farming methods and access to technology. In Pakistan, cotton is grown on 3 million hectares, mainly in the provinces of Punjab and Sindh. More than half of the farms are under 2 hectares in area, although they cover only 11 per cent of the area. However, fewer than 2 per cent of the farms covering 24 per cent of the area are larger than 20 hectares in size. Key actors in this segment of the chain are the 1.3 million farmers, 20 pesticide companies, 114 seed companies, government seed corporations, government seed certification department, the agricultural extension system, the cotton crop research institutes, the irrigation department, commission agents, ginners and agricultural credit companies. While some of these (such as the pesticide companies) are more organized than others, it is not clear whether there is a governance structure to influence inputs and outputs.

The processing stage covers activities involved in the transformation of cotton lint into cloth or garments for consumer use.[9] Textiles are a labour-intensive industry at the production stage, and employ proportionately more workers per unit of output than most comparable industrial sectors. The result is that economies of scale are not very significant, manufacturing is fairly competitive, and Southern manufacturers have an edge. However, the design and marketing stage is characterized by innovation-intensity, with considerable economies of scale and barriers to entry. This pattern corresponds closely to Pakistan's experience, where the majority of garment manufacturing and weaving units are in the small-scale, informal sector, although spinning and processing are done largely in medium- to large-scale integrated plants.[10]

An overview of the cotton commodity chain in Pakistan shows that one hectare of land at current expected yield levels produces 581 kilograms of lint cotton, 1,162 kilograms of cotton seeds, 500 kilograms of cotton yarn and 5,801 square metres of cloth. If the entire crop were processed to produce cloth, it would be able to cover 58 per cent of the production area. In 1991–92, when yield levels were 32 per cent higher (769 kilograms), the cloth output would have covered 76 per cent of the area. Potential yield levels, however, can rise even higher, to as much as double the previous peak, and thus the cloth would be able to cover more than 150 per cent of the land area producing the cotton.

The calculations also used existing prices to calculate the value added at various stages of the production chain. Under these assumptions, a hectare of land produced Rs. 36,312 worth of cotton lint and Rs. 8,715 worth of cottonseeds. If the cotton is processed, this will result in Rs. 46,200 of yarn, Rs. 104,418 of grey cloth, or Rs. 174,030 of finished cloth. The value added at subsequent points in the chain is likely to be even higher. In some market segments the exporters of cloth or apparel receive only one-third of the market value of the final product, while over 50 per cent accrues to the retailers (Gereffi 1994).

Another way of looking at the figures is that of the Rs. 174,030 value of the finished cloth produced by one hectare of cotton, only Rs. 16,286 (or 9.36 per cent) accrues to the farmer as income, while another Rs. 18,914 (or 10.87 per cent) is allocated as costs of production. In fact, the bulk of the rents are captured beyond the finishing stage – in the production of garments and retail sales to customers in the North. In this sense, the transformation of cotton production, even if it involves a doubling of unit costs, is not a market problem, since it adds less than 10 per cent to the cost of the finished cloth and even less to the price of finished garments. There is evidence that the market is willing to provide as much as a 20 per cent mark-up on green cotton products.[11] In fact, the problem lies in the domain of governance and technology transfer rather than that of production costs and consumer preferences.

As mentioned earlier, the bulk of the production of cotton and cotton products is for export. According to the calculations given below, between two-thirds and three-quarters of the cotton produced is exported in one form or another. The result is that this sector constitutes the most significant component of foreign exchange earnings for the country. The most recent year for which these figures were available is

1995–96, when the exports of cotton and cotton products were as shown in Table 3.1.

The calculations in Table 3.1 are designed to bring out the relation between domestic production and exports at various stages of the production chain. Figures on cotton output and exports are available directly from government statistics, as are those of export of cotton products. These figures, as well as those for yarn output, are also available from the annual review report by the chairman, All Pakistan Textile Mills Association (APTMA). However, output data on finished cloth and other made-ups are somewhat confusing because of variable coverage. Most documents report data on cloth output only from the 503 units in the mill sector that report to the APTMA.

In the 1990s, cotton production reached a peak of 2.2 million tonnes in 1991–92 and a trough of 1.4 million tonnes in 1993–94. Of this, a variable amount, depending upon the residual from the domestic demand of the yarn industry, was exported and the bulk of the cotton, between 1.3 and 1.6 million tonnes, was used domestically to produce between 1.1 and 1.4 million tonnes of yarn. Of this, about 45 per cent, or 0.5 to 0.6 million tonnes, were exported and the rest used for producing cloth, towel, canvas and other cotton products. As per our calculations, 500 kilograms of yarn produce 5,801 square metres of finished cloth, and a proportionately smaller quantity of canvas or towel. If the residual yarn production of 0.6 to 0.8 million tonnes were used exclusively to produce finished cloth, aggregate national output would come to between 7,000 and 9,000 million square metres. Against

Table 3.1 Exports of cotton and cotton products, 1995–96

Item	Value (Rs. billion)	Percentage of sector value
Cotton	19.44	13.24
Cotton waste	0.17	0.11
Cotton yarn	54.06	36.84
Cotton cloth	43.28	29.50
Speciality items	2.14	1.46
Garments	27.64	18.84
Total cotton sector	146.73	100.00
Total exports	294.74	

Source: Government of Pakistan 1997b. The figures for garment exports appear to be incomplete, since they are significantly lower than those provided by the All Pakistan Textile Mills Association (APTMA).

this potential figure, the output of the organized mill sector is only 300 to 325 million square metres. However, the actual national production is much higher, since cloth exports alone range between 1,046 and 1,196 million square metres, canvas exports (in 1995–96) 90 million square metres (constituting over 90 per cent of output), and towel exports 400 million square metres.

Up to 25 per cent of cotton production, 45 per cent of yarn output and over 20 per cent of the potential output of cloth and made-ups is exported. Taking all this into account, it is fair to say that more than two-thirds (and possibly three-quarters) of the goods produced from the original crop is exported. In other words, out of a total cotton area of 2.7 to 3.0 million hectares, exports account for 0.15 to 0.59 million hectares in the form of cotton, over one million hectares in the form of yarn, and over 200,000 hectares in the form of finished cloth (see Table 3.2). The numbers corresponding to the export of canvas, towel, hosiery and garments cannot be estimated directly, but is likely to be higher than the area accounted for by cloth exports.

Another way of approaching this issue is by looking at the value of output produced at various stages, and comparing it to export values.[12] The value of the total cotton output from 2.997 million hectares at prices then prevailing is Rs. 108.83 billion, of which Rs. 19.44 billion was exported, leaving a remainder for domestic consumption of Rs. 89.39 billion. If the entire cotton crop had been processed to produce finished cloth, the value of this cloth would have been Rs. 521.57 billion. A few simple calculations will reveal that of this volume, a little less than one-third is consumed within the country and the rest exported.

Table 3.2 Calculations of area used for export of cotton products

Year	Total cotton		Cotton exports		Yarn exports		Cloth exports	
	qty (m. tonnes)	yield (m. tonnes)	qty (m. tonnes)	area (m. ha)	qty (m. kg)	area (m. ha)	qty (m²)	area (m. ha)
1990–91	1.638	615	0.283	0.460	501.1	0.948	1.056	0.172
1991–92	2.181	769	0.456	0.593	505.9	0.764	1.196	0.156
1992–93	1.540	543	0.265	0.488	555.3	1.188	1.127	0.208
1993–94	1.368	488	0.075	0.154	578.5	1.378	1.046	0.215
1994–95	1.479	558	–	–	522.1	1.087	1.160	0.208

Source: Cotton yields, area and exports, Government of Pakistan 1997a; yarn and cloth exports, APTMA 1996; conversion from quantities to land area by author.

The value of the yarn produced from this cotton was Rs. 114 billion. Of this, Rs. 54 billion was exported, leaving a residual of Rs. 60 billion. This volume of yarn would produce grey cloth worth Rs. 135 billion, and finished cloth worth Rs. 225 billion. Of this, a total of Rs. 45 billion was exported as cloth or speciality items, and 61 billion as garments and made-ups. Assuming that only finished cloth is exported, the cloth remaining in the country would be valued at Rs. 179 billion. This is used to produce garments or household furnishings. From these, Rs. 61 billion worth of ready-made garments, hosiery and made-ups are exported. Assuming a 40 per cent mark up on garment production, the value of cloth in the garment exports would be Rs. 44 billion, leaving Rs. 135 billion worth of cloth for domestic consumption. This comprises 26 per cent of the potential cloth output (if the entire crop had been processed at home). These calculations can be summarized as shown in Table 3.3. The calculation yields a similar estimate, that roughly 30 per cent of the products of the cotton crop are consumed at home and 70 per cent are exported.

The relationship of the exporters with international buyers (as well as domestic ones) is extremely varied. Whereas garment producers (and some larger cloth producers) enter into direct relationships with international buyers, cotton, yarn and cloth manufacturers sell their product to commission agents, who grade it and sell it to domestic consumers as well as foreign buyers.

Even among garment producers, relationships vary depending on the type of product. In general, there are three types of consumer product: high-end fashion products, standardized mass-produced products and non-standard informal sector products. The former are speciality products, not suitable for mass production. These are generally produced in Third World countries through licensing or other arrangements with brand name companies. Standardized products are produced in large-scale manufacturing units located in the North (for example, Levi Strauss and Co.), or through their licensing arrangements in which they ship

Table 3.3 Estimated share of domestic use and exports of cotton products

Segment	Production	Domestic use	Export
Cotton	109	89	20
Yarn	114	60	54
Finished cloth	225	179	46
Garments	251	61	190

US-made parts for sewing overseas. Since brand name managers have a degree of market power similar to that of mass retailers, their relationship with cloth suppliers is similarly unequal. Finally, at the lower end are non-branded, lower-quality and often non-standardized products manufactured by small-scale suppliers in the South, and imported by mid-level or low-end retailers. The last sell mainly in domestic markets, or to small-scale buyers in the North. International retailers do not affect their behaviour in the same way as they do the other two groups.

Within the country, there is a complex set of relationships between textile companies, yarn producers, ginners, farmers, suppliers of farm inputs (seeds, fertilizer and pesticides), credit companies, and traders or middlemen (called commission agents) who operate at every step of the chain. However, the absence of a monopolistic structure within the country is linked to an absence of a coherent governance structure in products where the domestic segment is dominant, namely cotton production. As a result, the input and output relationships of the farm to the textile mill are far less predictable than those between the garment manufacturers and importer.

In this regard, however, a significant role is played by collective organizations of producers as well as others that support or influence market conditions, including interest groups, industrial lobbies and regulating institutions. The All Pakistan Textile Mills Association (APTMA), which represents mainly the interests of large-scale spinning units, is widely considered to be the most powerful industrial lobby in the country. Given the size of the textile sector in the national economy, the Federation of Pakistan Chambers of Commerce and Industry (FPCCI), and some of its member chambers, especially the Faisalabad chamber, are also active promoters of textile interests. APTMA represents all textile manufacturers, but is dominated by yarn producers. A smaller body, the All Pakistan Textile Processing Mills Association (APTPMA) is not as influential, but is more closely connected to the upper end of the production chain.

Cotton Production

In this section, we look at the main features of cotton production in Pakistan, the broad trends in output and yields, environmental problems, technological alternatives, the system of governance and the underlying system of knowledge. The goal is to identify mechanisms through which optimal technological choices can be introduced and a change in production systems brought about.

Trends and descriptions Cotton is an oil crop, though grown mainly for its fibre. The fibre consists of long, fine, flattened and convoluted hairs called 'lint', which can be detached easily from the seed. The value and quality of the cotton variety depends on the fineness of the fibre as well as its length. The longer and finer the staple the better its quality, since it can be used to produce thinner and lighter textiles without knots or uneven surfaces. A single fibre is a little less in diameter than a human hair, and is measured in micronaires. Five different staple lengths are distinguished: short (<21 mm), medium (21–25 mm), medium long (26–28 mm), long (28–34 mm), and extra long (>35 mm). The majority of the world production (about 60 per cent) consists of medium long staple. Medium staple accounts for around 18 per cent, and short staple a mere 3 per cent, produced almost exclusively in South Asia. Longer staple lengths (long and extra long), comprise around 18 per cent of the world production of cotton (during 1977–78 to 1981–82), and can be grown only in more or less ideal conditions as regards soil, water, temperature and light.[13]

Cotton production and yields increased dramatically over the nineteenth century.[14] In 1834, total global production was estimated at 340,000 tonnes. By the end of the century, they had risen almost tenfold to 3 million tonnes and by 1924–25 to 4.4 million tonnes (of which 93 per cent was produced by five countries: the USA, India (including present-day Pakistan and Bangladesh), China, Egypt and Brazil. Global output in the 1990s varied between 18 and 21 million tonnes, a fourfold increase over the last half-century. Three-quarters of this increase is accounted for by increased yields, which rose from an average of 200 to close to 600 kilograms per hectare (see Table 3.4).

In Pakistan, the area under cotton increased almost at a constant rate, with minor fluctuations from a little over 1 million hectares at the time of independence in 1947 to 3.1 million hectares in 1996–97. Between independence and the peak year of 1991–92, total cotton output increased more than elevenfold, from 0.2 to 2.2 million tonnes. The most dramatic expansion took place in the 1980s, when output tripled from an annual average of slightly over 0.7 million tonnes during the four-year period of 1979–82 to 2.2 million in 1991–92.

Slightly more than half of the increase in total output is accounted for by yield expansion. Yield trends can be divided into the following five different phases:

- *1950s: constant yields* In the 1950s, yields remained more or less constant for the entire decade, from 1949–50 to 1959–60, at around 200 kilograms per hectare.

- *1960s: steady growth* The first spurt of growth took place in the 1960s, when yields rose steadily from 200 to 300 kilograms per hectare in 1970–71, and to 361 kilograms in 1971–72.
- *1970s: the first cotton crisis* A severe and persistent attack of the American bollworm devastated the crop during the 1970s, resulting in wildly fluctuating yields between a high of 377 and a low of 233 kilograms, re-attaining the 1971–72 figure only in 1979–82.
- *1980s: rapid growth* There was a dramatic growth in yields in the 1980s from 364 kilograms per hectare in 1982–83 to 769 kilograms in 1991–92. This was also a period in which the major expansion in pesticide use took place.
- *1990s: the second cotton crisis* Repeating the experience of 20 years earlier, the peak achieved in 1991–92 was followed by another severe and persistent pest attack, this time of the leaf curl virus and its disease vector, the whitefly. Yields dropped dramatically from 769 to between 500 and 600 kilograms per hectare.

Both periods of crisis in the cotton crop were associated with pest attacks, and with the emergence of pest resistance after a period of growing pesticide use and consequent yield expansion.

Pesticide use in cotton production Traditionally, the major obstacles to the expansion of cotton yields have been the inadequacy of water and

Table 3.4 Average cotton output and yields, 1946–50 and 1991–95

Country	1946–50		1991–95	
	Output ('000 tonnes)	Yield (kg/ha)	Output ('000 tonnes)	Yield (kg/ha)
China	169.8	72.3	4,719.4	787.2
USA	2,609.4	303.3	3,670.5	747.6
Central Asia	519.9	357.0	2,320.0	784.6
India	543.6	105.8	2,192.8	290.1
Pakistan	181.9	148.5	1,859.8	652.7
Brazil	291.6	149.5	580.2	369.9
Egypt	311.4	563.2	328.2	866.0
Mexico	118.7	299.8	86.4	516.3
Other	570.1	123.8	3,361.2	516.3
Total	5,316.4	202.3	19,118.5	582.1

Source: IISD/WWF 1997: 8–9.

attack by insects. Overcoming these obstacles has been the central element of the strategy of increasing yields. However, this strategy is precisely the source of environmental degradation. The establishment of irrigation systems has often resulted in rising water tables, water-logging, salinization and water wastage. More importantly, while the use of chemicals to control pest incidence has dramatically increased yield levels, at least in the short term, it has also been the major contributor to environmental degradation as measured in terms of adverse effects on human health, soil and water quality, local biodiversity and ecological balance.

In this section, we will focus primarily on the environmental impact of pesticide use even though cotton production is associated with other environmental hazards as well, pertaining mainly to the use of chemical fertilizers and irrigation water. In Pakistan, while the increased use of fertilizer and harnessing of irrigation water have environmental con-sequences, these are not particular to cotton. Cotton requires a slightly higher quantum of fertilizer per hectare than other major crops (except sugarcane); the differences are not particularly striking. Cotton is grown on 15 per cent of the cultivable area, and consumes roughly 20 per cent of the 2.5 million nutrient tonnes of chemical fertilizer used in the country.

Similarly, although there are environmental costs associated with the excessive use of water, they do not appear to be exceptional for the cotton crop. Cotton is a highly water-intensive crop, requiring 24 inches of water per hectare per year. In Pakistan, as in many other countries, the entire cotton crop is on irrigated land, and this has often been used to justify the construction of irrigation systems. However, the average water use in cotton production does not currently exceed the average per hectare water offtake in the country. The average annual water offtake from the Indus, including groundwater exploitation, comes to approximately 110 million acre feet (44 million hectare feet) for 21 million hectares. This yields an average figure of 25 inches of water per hectare, which compares quite favourably with the figure for cotton. The irrigation system in Pakistan was developed in response to geo-political considerations, namely the need to share river waters with India after Independence. In that sense, the problems with irrigation have to be taken as a given, rather than as a consequence of cotton production.

To return to pesticides, their use goes back at least to the late nineteenth and early twentieth centuries, when cotton growers in the USA and Latin America began using sulphur and nicotine to limit pest populations.[15] In the early 1900s, the United States Department of

Agriculture (USDA) developed and promoted the use of calcium arsenate, whose production increased from 23 tonnes in 1918 to 20,000 tonnes in 1935. These chemicals were somewhat limited in impact because they had to remain on the plant in order to be effective, but the heavy rains that were fairly common in cotton growing areas made this difficult (Murray 1994).

A major breakthrough came with the discovery in 1939 of the most famous organochlorine, DDT (which earned its discoverer, Paul Müller, the Nobel Prize for medicine in 1948). Hailed as a miracle substance that would produce a total victory on the insect front, DDT was used with great success during the Second World War to protect combatants against malaria and typhus epidemics. Released for civilian purposes at the end of the war, it began to be used as a household disinfectant as well as an all-purpose insecticide. Its discovery was followed by the discovery of at least twenty-five other synthetic organochlorine compounds between 1945 and 1953, including BHC, chlordane, toxaphene, aldrin, dieldrin, endrin, heptachlor, parathion methyl parathion and tetraethyl paraphosphate. All these were used as pesticides for prophylactic uses as well as in response to pest outbreaks, as the then current agronomic wisdom aimed at the total elimination of pests. Pesticide sales jumped from $40 million in 1939 to $260 million in 1954, and cotton yields rose dramatically, by as much as 100 per cent in some areas.

By the 1960s, concerns began to emerge about the persistence of organochlorines in soil, water and the food chain, and the effect this had on human health and the environment. As a result, these substances were gradually phased out, and are today banned by almost all countries. In replacement, although the innovation process had begun earlier, chemical companies introduced less persistent (but often more toxic) alternatives, namely organophosphates (methamidophos, monocrotophos, methyl parathion, profenofos, methomyl), pyrethroids (cypermethrin, cyhalothrin, cyfluthrin, deltamethrin, biphenthrin) and carbamates (aldicarb, methomyl). These have replaced the organochlorines, although some illegal use of the latter persists because of lax supervision in some Southern countries.

Direct figures on the use of pesticides on cotton in Pakistan are not available. However, reasonably reliable estimates can be obtained from different sources. Since Pakistani farmers do not use herbicides or defoliants (since cotton is picked manually), the main source of concern is the use of insecticides. The Ministry of Agriculture in newspaper reports (cited in SDPI 1997) gave an estimate of 14,950 tonnes of pesticide (with an active component of 6,000 tonnes) in 1996–97. In another report, the Ministry (Government of Pakistan 1995) estimated that 65 per cent of the total pesticide use in the country was for cotton.[16] In 1993–94, total pesticide use was 20,279 tonnes, yielding a total for cotton of slightly over 13,000 tonnes. Finally, the experts surveyed for this study estimated that in 1996–97, cotton farmers used 3.5 to 4.5 litres of formulated pesticides. For the entire cotton-growing area (3 million hectares), this comes to between 10.5 and 13.5 million litres (which is equivalent roughly to between 13,000 and 15,000 tonnes).

Although the volume and rate of pesticide use are somewhat lower than in the high pesticide-using countries, there are indications that the adverse impacts are just as severe. Murray (1994) reports figures for Central America cotton producers of as high as 18 kilograms per hectare and as many as 20 to 30 sprays per season, as compared with between eight and 13 sprays per season in Pakistan. However, there are indications that the country may already be on the pesticide treadmill. For one thing, pesticide use is about twice the level recommended by cotton researchers and extension staff. Second, the area covered as well as the number of sprays and volume of pesticides has increased dramatically over the last 15 years. The Ministry of Food and Agriculture reported that, before 1983, only 5 to 10 per cent of the cotton-growing area in the Punjab was treated with pesticides; by 1991, this had increased to between 95 and 98 per cent (Government of Pakistan 1995). Plant protection departments also report an almost fivefold increase in total pesticide spraying from 2.8 million hectares in 1982–83 to 13.119 million hectares in 1992–93.[17]

Environmental effects of pesticide use   Chemical pesticides affect human health as well as biological diversity and surface and groundwater quality. Although the full impact on human health is difficult to measure, especially in Southern countries, the acute toxicity of the substances is not in question. In addition, the World Health Organization (WHO) has concluded that exposure to pesticides is probably carcinogenic as well as toxic. Some pesticides leave persistent residues in soil, groundwater and the food chain, thus exposing the human population to slow and cumulative poisoning. Various studies estimate the impact to be as high as 20,000 people killed and 3 million poisoned every year (see IISD/WWF 1997: 12–13).

In general, pesticide-related illnesses became serious as the earlier organochlorines (most notably DDT) became less effective and more problematic because of their long-term harmful consequences, and were replaced by more toxic organophosphates. Unlike the low levels of acute toxicity caused by the former, even brief oral or dermal exposure to the latter produces acute poisoning and even fatality (see Murray 1994: 44–

5). Cottonfield workers in Southern countries are most vulnerable because of lack of awareness of pesticide impact, lack of strict implementation of safety measures, lack of readily available running water and exposure to pesticide-contaminated water for drinking or cleaning. Other sources of pesticide contamination are through animal feed (seedcake), water contamination through run-offs or fine mist during spraying, improper use of empty pesticide containers for other purposes, and inadequate or illegal disposal of expired or unused pesticides.

Typical pesticide poisoning symptoms include stomach cramps, dizziness, vomiting and heavy sweating. However, since these symptoms are not exclusive to pesticide poisoning, the incidence of pesticide poisoning is likely to be underestimated. This is compounded by the absence of proper treatment and record-keeping facilities in Southern countries. Murray (1994) documents the case of the department of León in Nicaragua, where improvements in the record-keeping system led to the discovery of massive under-reporting of pesticide-induced illnesses in official records. The officially reported incidents of such illnesses increased from 200 in 1983 to 1,266 by 1987. The increase was considered more a product of improved reporting than an actual increase in the incidence of poisoning. This discovery led researchers to conclude that only 23 per cent of the cases occurring in the countryside appeared in official reports. The revised estimates of 4,777 to 10,343 poisonings for 1989 among a rural population of 300,000 is higher than the reported rate of malaria, and makes pesticide poisoning by far the leading cause of work-related illness and injury (Murray 1994: 47–8).

Pesticides also affect wildlife, domestic animals and biological diversity. Given the prevailing agronomic wisdom of the times, cotton farmers in the last half-century sought to transform the ecological system to eliminate insects altogether. The result of this all-out war was a severe reduction in soil quality and fertility (Murray 1994; IISD/WWF 1997; Edwards 1993; Dinham 1993). Other effects include mortality of birds and aquatic organisms. Although there are no precise estimates, there is considerable evidence of fish stock reduction in pesticide-intensive areas. Finally, human health and biological diversity are affected by contamination of surface water and groundwater due to agricultural run-offs.

Another consequence of pesticide use is what is termed the 'pesticide treadmill'. This refers to the situation whereby higher and higher doses of pesticides are required to control pest populations because of the development of resistance in pests and in the elimination of pest predators. An indication of this is the recurrence and persistence of pest attacks and volatility in yield statistics. Initially, the use of DDT and other organochlorines in the 1940s and 1950s increased yields dramatically for a decade or so. But this resulted in the development of pest resistance, which induced farmers to use larger and larger volumes of pesticides, thus increasing their costs, further damaging the ecological balance between pests and their natural enemies, and increasing pest resistance. The elimination of beneficial insects by excessive use meant that farmers could not afford to step off the treadmill because of anticipated crop losses. On top of this, the increasing demand for pesticides also led to increasing prices, thus raising costs even further.

This constitutes an economic argument for reducing or eliminating dependence on pesticides. Given that pesticides raise yields in the short term, only to lower them in the medium term along with an increase in costs due to excessive use, an argument could be made for the elimination of pesticides altogether. However, this option is not available to individual farmers, and requires a collective effort on the part of entire regions in order to be effective.

In Pakistan, there are indications of the emergence of the pesticide treadmill. The most direct evidence is the development of pest resistance to major chemicals. Recent laboratory experiments at the Central Cotton Research Institute (CCRI) in Multan reveal that the two major cotton pests, the American bollworm and the whitefly, have developed resistance against common pesticides. In particular, applications of monocrotophos had increased by 19 to 720 times to control these two pests, 26- to 168-fold increased applications of cypermethrin were required for the bollworm, and there was a 40- to 492-fold increase in methamidophos and a 104- to 725-fold increase in dimethoate use for the whitefly. Indirect evidence is provided by the long-term yield cycle. Yields increased significantly in the 1960s with the introduction of pesticides, but dropped suddenly with the attack of the American bollworm in the 1970s. This attack was overcome eventually because it was restricted to the deltapine variety of cotton; as the use of this variety was discontinued, the yield levels began to recover, and reached their earlier peak after a gap of ten years. Subsequently, in the 1980s, yields registered dramatic increases, only to collapse again because of the attack of the leaf curl virus and the whitefly. This attack has still not worked its way through the system.

The leaf curl virus was first observed in Pakistan in Multan in 1967 on a few plants (Hussain and Ali 1975), and occasionally thereafter until the first major outbreak in 1983 (Hussain and Mahmood 1988). The incidence increased dramatically after 1989 and reached epidemic proportions in 1992–93. The reasons for the outbreak are not fully known, but from similar experiences in other countries, it is reasonable to assume that the excessive use of pesticides was a major contributing factor.[18]

Very little work has been done on the pesticide cycle or on the effect of pesticides on non-target organisms. However, a number of studies have found pesticide residues in water and soil samples, seedcake, and people exposed to pesticide poisoning directly (such as cotton pickers). However, the studies do not appear to have been subjected to adequate screening and review. For example, Jabbar et al. (1993) found small concentrations of monocrotophos, cyhalothrin and endrin in groundwater samples taken from Samundri in the cotton-growing region. They also found traces of pesticides in soil samples. Although most of the concentrations were below the WHO limits, the indications are that persistent chemicals have deposited themselves in soil and groundwater. Similarly, a PARC study conducted in 1992 found that 71 per cent of the samples of cottonseed were contaminated with pesticides, including 39 per cent above the WHO maximum residue level (PARC 1992). Since 60 per cent of the edible oil in Pakistan is made from cottonseed, this constitutes a serious health hazard.

A more serious concern is that of pesticide poisoning, since there is no regular programme for monitoring this. The standard tests use blood acetylcholinestrase inhibition (AChE). A study conducted by the CCRI found that out of 88 female cotton pickers, only one had a sufficiently low AChE level (below 12.5 per cent), 74 per cent had moderate pesticide poisoning (AChE levels between 12.5 and 50 per cent) and one-quarter had dangerous AChE levels (50–87.5 per cent). The comparative figures for the 33 male cotton pickers were 12, 51 and 36 per cent respectively.

In countries such as Pakistan, another source of pesticide poisoning is the lack of information on pesticide handling. There are reported instances of obsolete and banned pesticides, lack of information or safeguards for cotton pickers, and improper disposal of pesticide containers. The Agricultural Pesticides Act 1972, which regulates the use of pesticides in the country, does not address the issue of safety standards and the lack of information for farmers as well as chemical industry workers.

Alternatives to conventional cotton While technological solutions that promise a more sustainable and environmentally friendly agriculture do exist, they have not yet been experimented with on a sufficiently large scale. These solutions include genetically engineered cotton, integrated pest management (IPM), integrated crop management (ICM), low external input sustainable agriculture (LEISA) and bio-dynamic agriculture, which requires cultural as well as production changes.

GENETICALLY ENGINEERED COTTON Work on the breeding of insect-resistant varieties is being undertaken at various research facilities. It includes modifying plant characteristics (such as the shape and size of leaves that form the source of nutrients for insects), the speed of ripening (thus limiting the exposure to insects during the vulnerable stage), and the introduction of insect-repellent genes into the plants. However, much of this is still in an experimental stage, and the economic viability as well as net environmental benefit has not yet been fully demonstrated (IISD/WWF 1997).

In Pakistan, since the attack of the leaf curl virus in the late 1980s, the focus of research has shifted to the development of varieties resistant to it. The varieties FH-682, MNH-147, BH-36 and CIM-240 have shown partial resistance to the virus in laboratory conditions. In fact, in 1995–96, the original crop estimate of 1.8 million tonnes was based on the assumption of virus-resistant varieties. The subsequent downward revision of the estimate (by 20 per cent) indicated that the degree of resistance was not adequate. However, this is a reactive strategy based on developing resistance after an outbreak of pest disease.

INTEGRATED PEST MANAGEMENT IPM is a common-sense method that builds on practices that farmers have used for centuries. For example, using varieties resistant to pests, altering times of sowing and harvesting, hoeing, removing crop residues and using botanical pesticides (such as neem and tobacco extracts). The name IPM goes back at least to the 1960s. In 1967, the FAO defined IPM as 'a pest management system that, in the context of the associated environment and the population dynamics of the pest species, utilises all suitable techniques in as compatible a manner as possible and maintains the pest population at levels below those causing economic injury'. It seeks to reduce pest populations to economically manageable levels through a combination of cultural control (such as crop rotation and intercropping), physical controls (hand-picking of pests, use of pheromones to trap pests), and less toxic chemical controls. However, it allows the use of chemical pesticides, even synthetic and toxic ones, when there is a need. IPM techniques are specific to the agro-ecological production conditions in any given location, and may involve the use of pesticides. As a result, few general principles can be applied and no absolute standards can be set for production.

In one study in India (Kishor 1992, cited in de Vries 1995: 46–7), it was found that the use of IPM resulted in higher yields and lower costs, resulting in a decline of unit costs by 28 per cent – although these are guessed estimates rather than precise estimates. Thus IPM use is found to be both economically efficient and environmentally beneficial. It also creates additional employment, since the technique is more labour-intensive than conventional methods.

A closely related alternative that is gaining ground is integrated crop management (ICM). As in the case of IPM, this alternative permits artificial fertilizers and pesticides when absolutely necessary, but aims to minimize them in favour of natural predators, crop rotation and hand-weeding. It differs from IPM, however, in that it goes beyond pest management to include issues of fertility, soil quality and crop management. Like IPM, it cannot be defined precisely, and has not yet become the basis of legally binding rules or practices.

While IPM is a promising alternative, and in some studies has been shown to be on the whole profitable for the farmers, it has not been very successful on a broader scale. Some reasons for its inability to replace conventional production may have to do with the fact that it involves more (and changing) work on the same area over time. Farmers might be reluctant to adopt a new concept because of a fear of unknown consequences. Also, it requires additional skilled labour, which might not be available. At the level of the enabling environment, reasons may have to do with the lack of an adequate service (i.e. research and extension) structure, uncertain initial costs of switching (because of a lack of received knowledge), and lack of region-specific knowledge of the pest complex. More importantly, it pits relatively low-paid and poorly motivated public officials and scientists against employees of powerful pesticide companies. Murray (1994) concludes his review of repeated promising but ultimately unsuccessful attempts to introduce IPM in Latin America by arguing that alternative approaches to pest control and sustainable agriculture need to be rooted in an alternative, participatory, egalitarian and small-scale agriculture, based on traditional knowledge. While I sympathize with the longer-term thrust of his remarks, in the short term, conventional approaches to the introduction of IPM do not offer much hope.

In Pakistan, research on IPM has a fairly long history. This was initiated as early as 1971 at the PARC research station in Rawalpindi, first as a seven-year PL-480 funded project on bollworms and a three-year PL-480 project on the whitefly, and an institutional support project on integrated pest management, funded by the Asian Development Bank.[19] However, these projects have not had a serious impact on production methods. A major reason is the limited nature of the project, without efforts to mainstream it in the functioning of the major research institutions, especially the system of cotton research institutes. Second, the extension system is not equipped to handle results from the IPM research, since there are hardly any avenues for training of extension staff in this technology. Finally, the educational institutions (especially the entomology departments of agriculture universities) do not provide the training or specialization in IPM necessary to ensure a steady stream of experts for staffing research, education and extension departments.

ORGANIC AGRICULTURE   One of the most environmentally friendly production methods that is gaining ground is organic agriculture, namely the production of cotton without the use of any chemical inputs whatsoever. Instead, it relies on natural processes to increase yields and disease resistance and enhance soil quality. At present, this is the only type that has an internationally recognized and independently assessed label for its products. Producers who do not meet the certification criteria but produce in an environmentally less harmful manner may not call their product organic, although they are allowed to use other terms such as clean, green or natural. In 1993, total global production of organic cotton was estimated to be between 6,000 and 8,000 tonnes, i.e. less than 0.04 per cent of the total global cotton output. Of this, about 75 per cent was in the USA and the rest in several countries, including 272 tonnes in India (in 1994–95). However, organic cotton is reported to be on a rising trend (by as much as 50 per cent in 1994).

In principle, organic cotton can be grown under the same conditions as conventional cotton. However, there are a number of caveats. First, it requires the farmer to make a considerable investment in studying the ecosystem of the area – although this could also be done by an effective and professional extension service. Second, current evidence suggests that organic methods will entail yield reductions of variable degree, which are not likely to be fully offset by the reduction in costs. Third, in some cases, the yield reductions might be so large, because of large natural population of pests, that organic methods might never be able to compete with conventionally grown cotton. Fourth, even otherwise, farmers will have to be prepared to meet initial losses because of a lack of balance in the production area; in particular, natural enemies of pests will not be abundantly available on first switching from conventional, chemical-dependent cotton. Finally, farmers will have to invest in crop certification, isolation of organic cotton from conventional produce, and cleaning of machinery and implements before use.

Although organic agriculture has simpler and more precise definitions and certification arrangements, there is nevertheless some variation across countries and regions, and at least one hundred regional or national standards have been developed so far. The EU and several national governments have developed framework legislation for organic agriculture. In the USA, a national law for organic products, the Organic Foods Production Act 1990, was revised and published in December 2000 along with the organic standards for production, processing and labelling, accreditation and importation. However, it will allow the label 'made with organic fibre' only for organically grown cotton, pending deliberation on the definition of organic textiles, which will include approved dyeing process standards.

The International Federation of Organic Agricultural Movements (IFOAM) has also developed international minimum standards. This is a worldwide umbrella organization of over 570 active member groups in 100 countries. Membership is open to associations of producers, traders and consultants as well as institutions involved in research and training committed to organic agriculture. IFOAM's accreditation system will ensure the existence of minimum levels for national or regional standards.

Until very recently, a three-year transition period from conventional to organic production was required for certification, presumably to ensure enough time for the reduction of pesticide and chemical residues in the soil. Cotton produced organically during the waiting period is described variously as 'transitional', 'pending certification' (California), or 'organic B' (Australia). However, in late 1994, IFOAM changed to a one-year transition period where levels of chemical inputs are lower.

Given the small size of the organic cotton sector, reliable estimates of longer-term productivity and economic trends are not fully available. De Vries (1995: 54–9) reports that the switch to organic cotton will involve an initial yield reduction of about 50 per cent, but that after some seasons, the net decline will be significantly smaller, probably around 30 per cent.

Some studies show yields rising to conventional levels, but this is rare, and is restricted to areas where insects did not play a large role in yield determination in the first place. The results from a micro-study of one such area, a Turkish cotton project, is cited by de Vries (1995).[20] This study (Zeegers 1993) shows seed cotton yields declining from 3,160 to 1,500 kg/ha in the short run (i.e. shortly after the transition), before stabilizing at 2,750 kg/ha by the time soil fertilization had improved. During this period, unit production costs rose from 100 to 211.4 before settling at 132.2, although later calculations for the same project are said to have yielded a unit cost escalation of only 11 per cent. In other words, the switch to organic cotton involves a unit cost increase of between 11 and 33 per cent, even in the long term, for an area with minimal insect presence. Other studies give conflicting results. In some cases, a unit cost escalation of between 50 and 150 per cent is reported (de Vries 1995: 58–9). An UNCTAD study on organic cotton production in India concludes that unit costs will rise by about 20 per cent, and suggests that the difference will narrow further with an expansion of the sector.

There is no recorded information on organic cotton production in Pakistan. However, at least three successful projects have been introduced in India, which has similar institutional and environmental conditions. These are small-scale projects, introduced with technical assistance from bilateral donors as well as marketing chains, in Madhya Pradesh, Gujarat and Maharashtra states. Technical assistance is provided by Swiss, Dutch and German institutions respectively. The assistance has included investment in extension services, access to credit and certification. Although these projects involve yield reductions, these are offset by the guaranteed price premiums of 20 per cent (Madhya Pradesh) and 10 per cent (Gujarat).

Given the small size of the market, organic cotton will face other costs as well. Foremost among these is the higher cost of spinning. This results from the fact that spinning machinery will have to be cleaned between running conventional and organic cotton. Also, given the small volume of the product, there might not be enough fibre of a particular length to make it feasible for dedicated spinning machinery, and the volume might not be sufficient to blend different fibres to produce consistent quality.

Data on the premium enjoyed by certified organic cotton are quite varied. Some studies indicate a doubling of price over conventionally grown varieties. The Indian studies mention a premium of only 20 per cent, although the hidden subsidies on extension and access to credit are also equivalent to an additional premium. It appears that while European consumers are prepared to pay the higher prices entailed in the switch to green cotton, the demand in the USA is fairly subdued.

To summarize the above discussion, there is sufficient evidence that technological alternatives do exist to enable a switch from conventional cotton production to more sustainable methods. However, to date, adoption has been limited. The question to be explored is that of mechanisms through which adoption can be enhanced. For this, we have to revert to the issue of governance.

System of governance As already mentioned, the 'governance system' refers to a set of arrangements and institutions that enable independent actors to make collective decisions in a legitimate and acceptable manner. To explore this concept further, in the case of cotton farming, we have to look at the manner in which farmers make allocative decisions, and, in particular, at how their decisions are affected by various interest groups. The first point to recognize is that farmers come into contact with three groups of actors who can influence them in making their production decisions. These are input suppliers (especially pesticide suppliers), purchasers (mainly commission agents), and government extension agents. Of these, only the government extension agents and input suppliers make an explicit effort to influence farmers' decisions. There is no history of commission agents trying to influence these decisions. Their role hitherto has been confined to purchasing cotton at market prices, sorting and classifying it, and selling it (generally after ginning) to spinners.

Let us begin with the issue of the farmers' incentive to modify production decisions. The simplest assumption is that this decision will be based on an assessment of the costs and benefits of various alternatives. In this scenario, farmers will switch to green alternatives if they become more profitable than conventional production practices. This can happen because of trade restrictions on conventionally produced output, because of premiums on green alternatives (such as eco-labelling), or finally because of awareness of the long-term adverse environmental and social consequences.

However, matters are rarely so simple. First, the uncertainty of crop yields means that the risk to farmers of a switch is going to be higher than that indicated by the aggregate figures. In other words, farmers will not only have to change production methods to unfamiliar ones, but will also have to assume the risks of yield variation. Second, the risks are likely to be even higher in the early years of transition, making it impossible for the demonstration effect of the innovation to influence lagging farmers to adopt the new technology. Third, the switch requires awareness of market opportunities and access to privileged markets. Since Pakistani cotton sells at a relatively low price on international markets, there is no evidence that farmers have tried to take advantage of high-value niche markets. It is not clear that the stimulus is there to take advantage of the green cotton niche. In any event, if it requires detailed technical assistance and extension by technical staff, the absence of such staff in the country means a lack of confidence in the ability to undertake the transition smoothly. Fourth, unlike the transition to chemical-based crop production in the 1960s, the success of the shift to green alternatives depends critically on the spread of the new technology. The larger the number of farmers adopting the new technology, the more productive will it become. This places a burden on the extension efforts far beyond what was experienced during the Green Revolution.

This last is a classic example of externalities faced by farmers in the form of the actions of other farmers. Classic economic theory bases its analysis of production decisions on the assumption of profit maximization. Given perfect competition, this yields completely independent decisions by farmers. However, even with these simple assumptions, the decision of each actor can depend critically and explicitly on those of others. In economic terms, this dependence is an 'externality', which can stem from market conditions,[21] scale economies in access to inputs (seeds, fertilizer, water, credit) and information (technology, learning by doing), and collective environmental impact (water and soil quality, the pest complex).[22] These externalities create incentives for collective input into decision-making.

In the traditional farming systems, such collective input tended to be highly localized, since the marketable surplus of most crops was fairly limited, and production was dominated by subsistence considerations. Cotton, however, has long been an exception to this pattern, and the expansion of output since the nineteenth century has been driven by export demand, initially in the British textile industry, and later in the global industry. However, this pattern is now found in all major crops, where the marketable surplus has grown rapidly since the advent of the Green Revolution in the 1960s. The result is a growing role of intervention in farming decisions by governmental research and extension departments.

However, governmental input and advice to farmers did not start in the 1960s. The colonial government initiated programmes of agricultural research and extension, albeit much smaller in scope and ambition, almost a century ago.[23] Initially, they focused on farmer education towards greater rationality and economic efficiency, better time management, the use of standardized weights and measures, and adoption of best practice or husbandry. The attention was on the process of production and attitudes towards production rather than on inputs as such. Seed production, manuring and pest control were still primarily in the hands of farmers rather than specialized organizations. In the 1950s and 1960s this began to change, as governmental agencies suddenly found a composite product to sell, and began to get involved more directly in decisions over inputs and technology. As a result, the amount and nature of scientific input into farming systems changed by a high order of magnitude in the post-independence and post-Second World War period. Yet the historical age of extension programmes was important in legitimizing the more broad-based intervention.

This 'composite product' was the combination of inputs and practices that goes by the name of the Green Revolution. Technologically, it was simply the introduction of high-yielding varieties of major crops, combined with 'optimal' application of fertilizers, water and other inputs. Besides raising per hectare yields to record levels, this also transformed the entire system and culture of agriculture production in the country. For the first time, it shifted the source of knowledge production from farmers to government bureaucrats and scientists. It privileged landlords over tenants, large and medium farmers over smaller ones, and governmental officials over farmers. It created a demand for transportation networks, markets, market intermediaries (commission agents), financial institutions and input-supplying corporations. It also aimed, somewhat less successfully, to reproduce in agriculture an 'industrial model' of operation, namely the creation of a controlled human and natural environment in which productivity could flourish.

It is important to emphasize three points. First, the Green Revolution took hold because it demonstrated success in raising farm revenues and profits. Second, it expanded 'organically' – in other words, it spread to a few farms, whose success led others to imitate them, and so on, until the entire process was completed in about ten years for wheat and rice, and 20 years for cotton. Third, it involved the earning of legitimacy and trust, to overcome a dissonance in the systems of knowledge between farmers and scientists.

In the initial years, the most striking successes were obtained in the wheat and rice sectors, based on research done at the CIMMYT in Mexico and the IRRI in the Philippines. While average cotton yields grew by 50 per cent in the decade of the 1960s, the major breakthrough in cotton arrived only in the 1980s, with a doubling of yields and tripling of output in ten years. Also, unlike the other crops, yield growth in cotton was far more difficult to sustain because of the changing nature of the pest complex.

As mentioned, the programme aimed not merely at the introduction of new or different inputs into the production process, but at the transformation of the process itself. It is not surprising, therefore, that it was led not by the invisible hand of the market but by the visible hand of a set of interlocking institutions– of agricultural policy, extension, research, credit, inputs (certified seeds, fertilizer, water) and education. Significant governmental and technical assistance resources were channelled into existing agricultural research institutions and additional ones were established. The governance of the research network was integrated and formalized through the establishment of the Pakistan Agricultural Research Council. New agricultural schools and colleges were established and older ones expanded and upgraded into universities. New government-owned corporations were set up to supply seeds (Pakistan Agriculture Seed and Storage Corporation – PASSCO), credit (Agricultural Development Bank of Pakistan – ADBP) and fertilizer (National Fertiliser Corporation – NFC). The agricultural extension network was expanded, and pamphlets, brochures and regular radio programmes on agricultural knowledge were produced.[24]

Today, the number of institutions that influence agricultural allocation decisions runs into scores, if not hundreds. The most prominent of these are the Pakistan Central Cotton Committee (PCCC), the system of Cotton Research Institutes (CRIs), the Pakistan Agricultural Research Council system, the Textile Commissioner, the Agricultural Prices Commission, the Federal Seed Certification Department (FCSD), various agricultural universities and the Export Promotion Bureau (EPB). The country has 4,000 agricultural scientists, 500 agricultural extension agents, and proportional numbers of officials in seed certification and supply, agricultural machinery provision, policy development and agricultural pricing and credit, all in the public sector.

While farmers do not have an organized association, they are represented heavily in national and provincial legislatures and on cabinets of ministers. Since cotton farmers are concentrated in Southern Punjab and upper Sindh, they are able to act in concert through an interesting union of ethnic and sub-nationalist grouping and economic interest. In recent years, a non-governmental organization, the Farmer Associates Pakistan (FAP), has been established as a lobby of the larger farmers. Most of the constituents (i.e. most large farmers) as well as the main leaders of this organization come from the cotton-growing area. It is not clear whether this group has had any influence beyond that exercised by the political leaders through their presence in national and provincial legislatures.

However, corporate lobbying in the cotton production sector does exist, not by farmers, but by input suppliers. The Pakistan Agricultural Pesticides Association (PAPA), is a powerful lobbying group, the bulk of whose activities are concentrated in the cotton system. Large seed corporations are also said to be involved in advocacy, lobbying and pressuring the government to deploy policies favourable to their interests.

To return to the Green Revolution, this system was driven by a simple philosophy: maximize yield. Researchers were directed to produce high-yielding varieties of crops, and extension agents mobilized to disseminate the information to farmers. Simultaneously, credit agencies were instructed to supply credit for expanded 'working capital' demand and input suppliers to ensure that the recommended inputs were in fact available. Behind all this was an orchestrated government policy aimed at subsidizing the cost of inputs and research and at ensuring stable output prices to maintain market stability.

The government used a combination of pricing policies, taxes and tariffs, and investment both to subsidize inputs in chemical-based agriculture and to create a stable market for its output. Beginning in the 1960s, explicit subsidies were introduced on fertilizers, pesticides and agricultural machinery, as well as on complementary inputs (especially irrigation water and agricultural credit). Initially, fertilizer and pesticides were supplied directly by government agencies at controlled prices, which were generally lower than the market prices. Pesticides in particular were made available virtually free of cost in the beginning, and agricultural self-sufficiency was closely linked to the amount of pesticide used. In 1986, a process of liberalization was started to bring input and output prices in line with international market conditions. Still, prices were held down by a combination of low tariffs (the entire demand for pesticides is met through imports), free technical input (through research and extension agencies) and liberal regulations. Gradually, the tariffs on pesticides were also brought in line with the overall tariff levels, and today equal about 27 per cent of import value.

This process also coincided with the granting of permission to the private sector to supply agricultural inputs. As a result, the private sector now has a free hand in the manufacture, import and marketing of pesticides. Extension agents of pesticide companies try constantly to induce farmers to use more pesticides, and farmers have no choice but to accept this advice since none other is available. On the one hand, crops have become susceptible to a broader spectrum of pests, and on the other hand, pesticides are becoming increasingly expensive due to the monopoly of multinational companies. Moreover, most of the key pests have developed resistance to pesticides and their larger populations pose an increasing threat despite the ever-increasing usage of pesticides every year.

This situation contains hidden subsidies for pesticides, in the form of inadequate information to farmers on the comparative costs of pesticides and their alternatives. Farmers lack knowledge on the side-effects of pesticides, and governments are loath to provide this knowledge, presumably out of fear that it may obstruct the goal of maximizing yield. This proves harmful to agricultural productivity in the long term. As pest numbers and population increased because of excessive use, a switch to other policies and practices was warranted but did not take place. Instead, there is a tendency to increase the intensity of use to compensate for the loss of effectiveness.

In retrospect, if one ignores the environmental and social costs and focuses on the yield levels alone, the Green Revolution programme could claim to have been extremely successful. It led both to the introduction of new inputs and to the replacement of an existing system of knowledge with another one. Today, high-yielding varieties of wheat, rice and cotton cover virtually the entire cultivated area (although as many as 50 per cent of the seeds used are not 'certified'). In the meantime, the use of chemical fertilizers has grown at a rate of about 7 per cent per year, and the application of chemical pesticides from a minuscule amount in the 1950s to over 20,000 tonnes in the 1990s (now amounting to 15 per cent of total production costs in cotton). Supporting these initiatives is the irrigation network, which has more than doubled in reach.

The result is a rapid transformation of agriculture as a cultural and social activity. First, the landlord–tenant relationship came under severe stress in the 1960s and again in the 1980s, leading to peasant uprisings, armed conflict and periods of lawlessness and open civic unrest in the rural areas. Although other factors also contributed to the emergence of unrest,[25] the most significant is the Green Revolution. On the one hand, it raised incomes, thus requiring a rewriting of the traditional agreements for the sharing of the benefits. On the other hand, it undermined peasants' knowledge, which was the basis of their bargaining power with the landlords, and replaced it with scientists' knowledge to which the landlords had preferential access.

Second, the collective knowledge of traditional agricultural practice, which is viewed by many scholars as more environmentally benign, has more or less died out over vast areas, and has been replaced with a growing dependence on external institutions and experts. This is especially the case with regard to crops vulnerable to pest attacks, such as cotton, fruit and vegetables. Initially, the external institutions were strong and well supported financially and politically. Over time, they have deteriorated in quality or have been captured by powerful interest groups. The responsibility of some institutions (such as PASSCO, the NFC and the Cotton Export Corporation) has shrunk in proportion to the expansion in the private sector. Others have been debilitated by years of mismanagement and corruption (for instance, the ADBP). All organizations are threatened by the generalized crisis of governance in the country, and declining salary levels and poor monitoring systems have led to a dramatic deterioration of the efficiency, integrity and transparency of government organizations. This is especially true in areas (such as research or extension) where output cannot be monitored directly. Finally, a number of functions are under threat of capture by specialized interest groups. Foremost amongst these is the pesticide supply system.

Notwithstanding its growing weakness, the governmental system of agricultural support functions in a manner that is strikingly similar to and perhaps more ambitious than the role ascribed by Gereffi (1994) to large retail corporations in buyer-driven commodity chains. The government behaves as a large corporation whose goal is to maximize the output produced by its constituents, and thus to maximize the total earnings for the country from this output. Unlike retail corporations, the stake of the governmental system is not directly pecuniary, nor has it expanded disproportionately with the increase in the total value of output. Yet the government does have a stake in this output, given that a higher agricultural output means higher GDP, exports, imports and tax revenues, greater political stability (at least in urban areas) and thus less demand on resources in the hands of the state. This is both a pecuniary and a political interest.

The similarities end here. Unlike retail corporations, however, whose goal is the maximization of profits, the goal of the governmental system is the maximization of yields. Thus, while the former adapted to changing market conditions, the latter had the freedom of being more conservative. Actions that enhance *net* output (after deducting all social and environmental costs) could be forced upon a corporation by market considerations, but can be ignored by the government if they go against the grain of the explicit national goal. This is certainly the case with sophisticated interventions (such as IPM) that are supposed to raise net profitability even though they depress yields and output. Although a small number of agricultural scientists have studied this issue, and have made considerable progress in selected areas, their research has not been disseminated properly, nor its significance to farmers demonstrated properly.

Second, as mentioned already, governmental action tends to become captive to strong vested interests. For example, the supply of pesticides is in the hands of powerful multinational corporations, who employ highly paid researchers and extension agents (sales representatives) for their products, and who are also believed to influence governmental policy in their favour. IPM and organic cotton do not have a similar profit motive backing them.

A major stumbling-block is the very size and inertia of the agricultural establishment. The vast majority of researchers and extension agents are schooled in the old-style 'maximize yield' philosophy. Switching to IPM involves a significant personal investment by these individuals, for which there is no institutional or other incentive. In contrast, sales representatives of private companies (dealing with pesticides, seed, fertilizer, agricultural machinery) have an incentive to keep up with the research to promote their products (and therefore their careers).

However, of all the agencies that have a direct personal contact with farmers and try to influence the farmers' allocative decisions, the government research and extension is the only one that has the potential of influencing the switch to cleaner production. Private pesticide and fertilizer companies face precisely the reverse incentives, namely to maintain and even intensify the current production practices in order to increase the demand for their products. The purchasers of cotton, who might be viewed as having a private stake in improved varieties, have neither a history of interaction and extension with farmers nor one of responding to new market niches. Also, they are not organized collectively, and lack a sufficiently concentrated structure to enable leading firms to act as innovators and price leaders.

In other words, of the four groups, two (the farmers and the purchasers) do not exhibit a corporate structure or mentality, making it difficult for their collective interests to find expression or to be enforced. A third group (the input suppliers) recognize their interest clearly and enforce it effectively through advertising, extension and research, but this interest is contrary to the logic of sustainable agriculture. The fourth group (the government) interprets its collective interests to lie in the maximization of yields, and pursues this objective with a certain degree of historical inertia, if not with great efficacy. However, since this does represent the public interest, including the interest of farmers, it is quite conceivable that its interpretation of the collective interest may change. That will raise the issue of whether it has the institutional strength and political will to pursue the collective interest in the face of opposition from vested interests.

A feasible programme of transition to sustainable agriculture will therefore involve the following characteristics:

- The transition cannot be based on market incentives alone, since there is no technical and informational link between final consumer markets and farmers.
- Given the low-level market niche of Pakistani cotton, and the enormous size of the cotton output, it is unlikely that the transition can be led by the lure of high-end market niches. It is therefore better to focus on medium or low-end cotton production as the goal. This can mean a concentration, at least in the first stage, on IPM rather than organic cotton as the goal of the transition.
- Since definitions of IPM are even looser than those of organic cotton,

they cannot be readily converted into advice to farmers. There is a need to start a process for defining IPM more precisely.

- The transition to IPM must involve the research and extension system of the government in a central manner. This cannot be done through marginal investment in IPM research.
- This involvement can be obtained only if it is in the interest of the research establishment to undertake IPM activities. This can be done through the establishment of career paths, research and training opportunities, international exposure, and even commercial opportunities for researchers.
- Similarly, there must be clear incentives for extension officials and university faculties to switch to IPM-based education and practice.
- Finally, the transition must involve pesticide companies in a central manner. This can be done on the lines of the business–government roundtable introduced to clean up the textile processing industry (see below). It may involve providing incentives to chemical companies to introduce IPM technology.
- Finally, there is a need to establish crop insurance programmes to reduce risk to farmers during the transition.

Textiles

Introduction[26] The textile industry is generally sub-divided into three sectors, namely spinning, weaving, and composites or processing. Lint cotton, separated from seeds, is fed into spinning mills, which produce yarn exclusively. The yarn is supplied to weaving mills, which produce grey cloth exclusively. Finally, grey cloth is supplied to processing mills, which produce finished cloth. Alternatively, lint and yarn are supplied directly to composites, which are vertically integrated facilities that produce a variety of products ranging from yarn to printed and finished and dyed products. A fourth sector that has taken off in recent years is that of apparel and knitwear, which produces ready-made garments, hosiery and other made-up products. Cotton processing is a highly polluting industry, utilizing a number of chemicals in desizing, mercerizing, bleaching, dyeing and finishing of cloth.

A good description of the processing stages can be found in IISD/WWF (1997) or SDPI (1995). Briefly, the spinning stage consists of blowing and mixing, carding, combing, drawing, simplex, ring spinning and cone winding. The weaving stage comprises warping, sizing and weaving. Finally, the processing stage covers singeing (burning to remove loose threads), desizing (removing a flexible film, called 'sizes', from the fabric), scouring (immersion in caustic soda to remove impurities), bleaching (using chlorine, hydrogen peroxide or sodium hypochloride to whiten it), heating, washing, drying, mercerizing (immersion in cold caustic soda to improve affinity to dyes), washing, dyeing, washing, drying, printing, drying and finishing (e.g. to obtain wrinkle-free or water-repellent clothing), and calendering. Each step involves environmental costs because of the characteristics of the effluent discharged. However, methods are available for minimizing and even eliminating the environmental costs of the processes. The finished cloth is used further along the chain, in garment manufacture.

The various steps in the production chain correspond to particular industrial structures. The yarn-producing sector comprises mainly large-scale units. These are organized collectively under the auspices of Pakistan's most powerful industrial lobby, the All Pakistan Textile Mills Association (APTMA). Yarn exports go mainly to Asian markets in large volumes. The APTMA's strength translates into considerable influence on the government's policies on prices, international trade, and credit in favour of the spinning sector.[27] The net result of these policies is a high degree of protection to the yarn sector through low cotton prices and high yarn prices. This protection has nurtured a relatively inefficient industry, which has not been able to graduate beyond coarse-quality yarn after 40 years of development. Eighty per cent of the total yarn production in the country still comes into this category. The result is that Pakistan has remained an exporter of low-quality fabrics and garments.

The weaving sector is mainly an informal sector industry, without adequate collective representation. However, 10 per cent of the weaving units are in the organized 'mill' sector, and are also influential in the APTMA. Yet the weight of the industrial lobby is in the spinning sector. Cloth goes mainly to European and American markets, and are bought both by large importers and by garment-making operations (such as Levi Strauss and Co.).

The textile-processing and composite units are organized under the All Pakistan Textile Processing Mills Association (APTPMA), a smaller body with considerably less clout in policy circles. Similarly, garment manufacturers have their own collective association, the Pakistan Cotton Fashion Apparel Manufacturers and Exporters Association (PCFAMEA), which also has less influence. These garment manufacturers are generally tied to international retail houses. Their goal is to obtain long-term supply contracts with large retailers.

In Pakistan, at the time of independence in 1947, despite the existence of thriving cotton farming, there was very little industrial activity using cotton. Only six spinning factories, 80,000 spindles and 3,000 weaving machines processed cotton, supplying about 8 per cent of the domestic demand. Today, the textiles sector accounts for about 19 per cent of value added in manufacturing, 64 per cent of exports and 40 per cent of manufacturing employment. It is by far the largest industrial sector in the country, out of a much larger industrial sector. Since 1947, the share of value added in industry has risen from a negligible level to over 27 per cent, exceeding the share contributed by agriculture.

The installed capacity in the formal, large-scale yarn-producing sector alone has increased to 8.7 million spindles, 143,000 rotors, 14,000 looms, and 1,312 shuttle-less looms, in 503 production units. Beyond this stage, the textile sector comprises 650 textile-processing units (including some state-of-the-art machinery), about 10,000 knitting machines in 600 knit-wear units (including 80 fully integrated firms), 6,500 towel looms, 225,253 weaving units (mainly in the informal sector) and 1,000 garment manufacturers.

As in many Southern countries, textiles has been the engine of growth for Pakistan's economy. The industrialization process in the country started with yarn production, from where it expanded into weaving, dyeing and finishing, and most recently into made-ups, hosiery and garments. Yarn production increased from 0.43 million tonnes in 1980–81 to 1.37 million tonnes in 1993–94. During this time, cloth production in the mill sector was stagnant, ranging between 325 million square metres in 1980–81 to 322 million square metres in 1993–94. However, the value of exports of yarn and cloth grew at similar rates between 1982–83 and 1995–96, from Rs. 3.56 billion and Rs. 3.62 billion to Rs. 54.08 billion and Rs. 45.42 billion respectively.

While similar output figures for garments are not available, export values increased dramatically in the 1980s. In 1994–95, the combined foreign exchange earnings from the last stage of production was Rs. 61 billion, compared to Rs. 49.57 billion from yarn exports, Rs. 34.99 billion from cloth exports, and Rs. 4.06 billion from cotton exports. Nevertheless, the full potential from this sector has not yet been achieved. A major source of retardation is the slow growth in Pakistan's textile quotas under the MFA.

The textile industry is the largest source of foreign exchange earnings in the country. In 1994–95, for example, yarn contributed 18.8 per cent of total exports, cloth 13.3 per cent, hosiery 8.5 per cent, cotton made-ups 8.0 per cent, and ready-made garments 7.9 per cent. This yields a total of 56.5 per cent from the export of processed cotton products. The major markets for cotton yarn are Japan, which imported $362 million in 1994–95, Hong Kong ($310 million), South Korea ($169 million), Dubai ($46 million) and Canada ($33 million). Cloth exports went mainly to Hong Kong ($163 million), the USA ($111 million),

Bangladesh ($80 million), Australia ($57 million), the UK ($51 million), Belgium ($50 million), South Korea ($49 million), Dubai ($45 million) and Japan ($42 million).

The market for cotton products is influenced by supply as well as demand factors. On the supply side, the main features of the textile industry in Pakistan are as follows:

- *Size of informal sector* An important feature of this sector is its non-mill component, consisting of independent, small-scale weaving units, which account for as much as 90 per cent of all fabric production. While the small size of the individual units provides an advantage in terms of greater flexibility, it also means poor management, weak financial capacity, and poor and outdated machinery.
- *The mill sector* Approximately 40 per cent of the members of the APTMA are integrated mills, namely mills that perform spinning and weaving and in some cases processing as well. In 1997, these mills had a total of 14,000 shuttle looms and 1,312 shuttle-less looms. In addition to this, the informal, non-mill sector consists of independent, small-scale weaving units and power looms. In 1997, there were 13,340 independent weaving units and 202,000 power looms. As a result, roughly 90 per cent of the weaving capacity is in the small-scale sector.
- *Integrated sector* In recent years, the country has also acquired state-of-the-art weaving machinery with some facilities for procuring modern dyeing and printing equipment. There are approximately 650 firms in this sector, with an installed capacity for producing 1,150 million square metres of finished cloth every year.
- *Knitwear sector* This sector comprises about 600 units, housing 10,000 knitting machines. Eighty of the firms are major integrated units with knitting, dyeing and sewing processes, while 60 per cent of the output is in the small-scale sector.
- *Garment manufacturers* The garment manufacturers number about 1,000. Most of them are small-scale household units, with fewer than 50 machines each. Of the 550 firms that are members of the PCFAMEA, 250 are small-scale firms (i.e. fewer than 50 sewing machines) and 300 are bigger enterprises (with 50 to 300 machines each).

The demand side is influenced by international as well as domestic considerations. On the domestic front, an estimated population growth rate of over 3 per cent per year, and a growth of income per capita of 2–4 per cent per year, translates into a demand growth of 3.5–7 per cent per year for cotton products. While figures for income elasticity of demand are not available for Pakistan, various estimates suggest a figure between 0.5 and 1.0 for low-income countries, probably closer to the lower figure. This means that demand will double over the next 20 years. If yield expansion can keep up with the previous trends, output will at least double during this period, yielding a significant surplus for exports. Even if yields remain stagnant, the share of exports in the total cotton crop will decline from its current 70 per cent to roughly 40 per cent by 2015. However, this might be accompanied by a shift towards higher-end products.

A major factor in influencing demand conditions is the recent GATT Agreement on Textiles and Clothing (ATC). Under this agreement, the trade restrictions under the Multi-Fibre Arrangement (MFA) would be phased out over a period of ten years starting in 1995. Trade restrictions on cotton products by Northern countries have a long history, going back to 1925. The goal of these restrictions was to protect declining Northern manufacturing industries from competition with lower-cost producers of textiles in Southern countries. The first international agreement on cotton textiles was negotiated in 1961 at the behest of the USA. Under the agreement signed in October 1962, quotas were imposed on cotton exports from Southern to Northern countries for a period of ten years. It allowed bilateral negotiations over quotas, and in some cases even unilateral imposition of quotas.

The MFA, a more elaborate agreement, was signed in 1974. Its scope was expanded both functionally (to include synthetic fibres) and geographically (to include newly industrializing countries). Successive rounds of agreement made the quotas even tighter, and expanded coverage to include other fibres. Since the 1980s, the growth rates of Pakistan's quotas have been reduced to below their target level of 6 per cent per year, and additional products have been brought under the purview of the restrictions. Since 1985, Pakistan's quota utilization rates have been extremely high, ranging on average between 82 and 96 per cent for the USA and (except for one year) between 105 and 119 per cent for the European Union. Item-wise, quotas have also been high, at 100 per cent or higher in five categories and over 80 per cent in twelve categories for the USA. This suggests that the MFA is a binding constraint on Pakistan's exports, and that the elimination of the quotas will provide a stimulus to exports.[28] However, given the nature of international competitiveness in this sector, the eventual results cannot be predicted with precision (Ingco and Winters 1995; Low 1995).

In recent years, the textile industry has gone through a series of crises. A major contributing factor is the decline in cotton yields and output after the peak year of 1991–92, coinciding with an increase in international cotton prices and a reduction in the global cotton crop in 1993. However, this is not the only factor: plant and machinery obsolescence, and inefficient management protected by years of favourable policies, have also contributed to the crisis. In any event, the increase in domestic cotton prices has been quite significant. Coupled with the scarcity of raw cotton, it led to a decline in the dollar value of exports in 1992–93 and 1993–94. From around Rs. 10.62/lb in 1991–92, the price of domestic cotton rose to Rs. 27.39/lb in 1994–95, an increase of 158 per cent in three years. This compares to a world price increase of 46 per cent in the same period, and a general wholesale price increase of 45 per cent (see Table 3.5).

The crisis in the textiles sector can also be discerned from the movement of share prices. In the 1980s, the stock market was gradually liberalized and expanded, with a big push coming in 1990–91. This process accompanied a booming market throughout the 1980s, and a dramatic rise in 1990–91. Share prices of cotton stocks tripled in 1989–90, leading to a generalized expansion of the share price index. However, the collapse of the cotton crop and political turbulence in 1992–93 resulted in a market crash, led again by a 50 per cent decline in cotton prices. The restoration of political stability produced a temporary revival in 1993–94, but then prices fell, and the cotton share price index consistently fell below the general index.[29]

Environmental problems As in the case of cotton production, the textile industry is associated with a number of environmental problems. The main source of pollution is the discharge of untreated effluents into

Table 3.5 Price increase in cotton

Year	Cotton (Rs./40 kg)			WPI	Cotton WPI
	K-68	NIAB-78	MNH-93	(1990–91 =100)	(1990–91 =100)
1991–92	933.10	908.36	951.59	109.84	106.04
1992–93	1,052.10	1,001.20	1060.38	117.92	119.20
1993–94	1,688.60	1,625.13	1722.72	137.26	168.20
1994–95	2,253.57	2,060.82	2,290.29	159.22	207.62
1995–96	2,221.20	2,113.04	2,245.84	176.90	210.57
1996–97	2,685.61	2,536.13	2,688.95	199.92	242.89

Note: WPI, K-68, NIAB-78 and MNH-93 are different cotton varieties
Source: Government of Pakistan (1997b).

water bodies and soils. Liquid effluents from washing, dyeing and bleaching operations contain organic and inorganic chemicals, as well as suspended solids (such as fibre and grease). Effluents are generally hot, alkaline, smelly and coloured by chemicals used in the dyeing process. Some chemicals are toxic. The effluents reduce dissolved oxygen levels in receiving water bodies, threaten aquatic life and damage the aesthetic value as well as the quality of water downstream.

We can distinguish between two types of environmental costs: those that affect the health of the final consumer and those that affect the local environment in the production process. Particular persistent organic chemicals used in the production process generally cause the former effect. Recent attention has focused on azo dyes, which are favoured by textile manufacturers because of their brilliant colours and qualities of adhesiveness. These dyes are believed to cause cancer, and have been banned by a number of European countries. However, information has not quite filtered down to producers, and laboratory tests in receiving countries reveal a high proportion of fabrics made with azo dyes.

The German government has banned the import of any products that may be in contact with the skin for prolonged periods – such as garments (including outer garments) – if they contain traces of azo dyes. Contravention of this legislation is treated as a criminal offence, and is subject to severe penalties including the burning of seized shipments at the manufacturer's expense. This legislation was contested in the European Court of Justice as constituting a trade barrier. However, the ban was upheld by the Court, and has led to similar legislation being passed in France, Sweden and the Netherlands. However, random tests conducted by German authorities on textile imports revealed that 48 per cent of Chinese, 34 per cent of Pakistani, 22 per cent of Indonesian, 19 per cent of Italian and 13 per cent of Indian exports contained the banned azo dyes and amines.

In the future, European importers are likely to require exporters to test shipments against azo contamination before exporting. Expressions of resentment against these measures have surfaced in the industry on the suspicion that they are meant to protect not final consumers but the uncompetitive European industry as the earlier protection provided by the MFA is phased out. In any case, efforts have begun to comply with the new laws. The Pakistan Export Promotion Bureau has held seminars and workshops to apprise producers of the problem and its solutions, and plans are afoot to establish a state-of-the-art testing facility in Lahore.

The prospects of product and process measures are quite different. On the one hand, there is evidence of successful use of such measures by US retailers in the case of labour standards, which are conceptually analogous to environmental standards. On the other hand, this experience is limited to the garments industry, and does not cover the full range of activities. Its application in other industrial sectors (carpets, leather and surgical instruments) has been less effective. We turn now to the treatment options in textile processing. A description of the major environmental concerns can be found in Chapter 4.

Treatment options The treatment recommendations presented here are derived from a roundtable process involving representatives from industry, labour, environmental NGOs, government regulatory and financial institutions, technical experts, environmental NGOs, technology suppliers and journalists. To this end, a series of detailed environmental audits of selected industrial units was conducted. The results of the audits were presented to the business–government roundtable process initiated under the national conservation strategy of Pakistan. The roundtable process involved over one hundred individuals from the functional groups listed above.

The roundtable participants were presented with the results of the environmental audits and with recommendations for action by government agencies as well as industrial units. A set of consensus decisions was finalized and submitted to the Pakistan Environmental Protection Council, the highest policy-making body on environmental issues in the country. The Council approved the recommendations, and constituted two standing committees to work out the details. These are the Committee on National Environmental Quality Standards (NEQS) and the Standing Committee on Technical Matters. The committees worked out the details for implementation of NEQS in consultation with industry representatives, and submitted their final recommendations. Some elements of these recommendations were subsequently enacted as part of a new Environmental Protection Act. Others were issued through executive orders. The recommendations fell into three groups: financial incentives for pollution abatement, the imposition of a pollution charge on effluent that does not meet quality standards, and the initiation of a proper system for monitoring and assessment (for more details, see Chapter 4).

According to the SDPI study, a switch to cleaner production would require the introduction of in-plant control mechanisms as well as end-of-pipe treatment. The latter would be more significant in existing industrial units. The treatment of industrial effluent can be classified into primary, secondary (or biological) and tertiary (or advanced physico-chemical) processes.

Primary treatment includes processes such as screening, neutralization, aeration, equalization and gravity sedimentation. The purpose of primary treatment is to remove suspended matter (including oil and grease) and to achieve uniform flows and concentrations. As the suspended matter is removed, BOD and COD levels are also reduced. An important component of primary treatment is the use of chemicals to neutralize the effect of other chemicals. Sulphuric acid is used to bring down the pH level of the effluent, and oxidizing chemicals (chlorine, sodium hypochlorite, calcium hypochlorite and ozone) are used to reduce colour through the oxidation of dye molecules. Similarly, lime, ferric chloride, alum and ferrous sulphate are used to react with dyes and form coagulants, which settle down in the treatment bed.

Secondary or biological treatment involves the development and cultivation of micro-organisms on food or substrate available in the effluent to lower its BOD. The process can be aerobic (i.e. in the presence of oxygen) or anaerobic. The most popular method is activated sludge treatment. It consists of a primary sedimentation tank, an aeration tank and a secondary sedimentation tank in a series. Provision is made to recycle settled biological sludge from the under-flow of the secondary sedimentation unit into the aeration tank to maintain the desired level of microbial population. Generally, low-energy options are preferred because of lower operating costs and higher reliability in Pakistan's institutional environment. Primary and secondary treatment of textile effluent can reduce BOD levels by 94 per cent and COD levels by 80 per cent.

Tertiary treatment involves a full chemical recovery of the effluent contents. It uses a high-rate multi-layer granular filtration with flocculation. The system consists of two pressure filters in series (double filtration) and the addition of chlorine, alum and polymeric flocculation. Each granular layer has three strata consisting of anthracite, quartz sand and garnet, which differ in density and granule size. Tertiary treatment can reduce BOD by 98 per cent and COD by 90 per cent.

Various in-plant control measures can substantially reduce the generation of wastewater, and thus reduce treatment costs. These measures include shutting off water supply to equipment not in use, installation of automatic shut-off valves on water lines, avoidance of spillage and preparation of only the required amount of chemical solutions. Integrated Pakistani manufacturers use many of these measures.

The roundtable process recommended that in the initial stage, industry be made to focus on reducing BOD, COD and TSS to within allowable limits. This could be accomplished by primary and secondary treatment alone. It is estimated that tertiary treatment and full chemical recovery would involve a capital cost of Rs. 160 million. In comparison, a conservative estimate of the cost of primary and secondary treatment for an average (1,500 tonnes per day) integrated textile-processing unit comes to Rs. 38.75 million in 1995. Of this, Rs. 22.65 million is for the purchase of plant and equipment (including imports of Rs. 8.3 million), Rs. 9.90 million for local costs for civil works and installation, and Rs. 6.20 million for interest costs during construction. In addition, the plant would require annual operating costs of Rs. 3.14 million. The annualization of capital costs depends on assumptions regarding interest costs, import duties and other taxes. These will range from between 0 (in the case of an outright grant) to Rs. 7.28 million per year. This yields a total annual cost ranging between Rs. 3.14 million and Rs. 10.42 million. To obtain a perspective on these figures, note that such a plant produces approximately 20 million square metres of cloth per year. In other words, the annual operating costs of treatment facilities are between Rs. 0.16 and Rs. 0.52 per square metre of finished cloth. Given an average sale price of Rs. 30 per square metre, total revenues are Rs. 600 million per year, and total annual treatment outlays (including capital cost) constitute between 0.5 and 1.5 per cent of the revenues.

The roundtable proposal recommended a pollution charge of up to Rs. 0.50 per square metre of finished cloth on polluting industries, and a series of financial incentives to induce voluntary compliance with environmental standards.

System of governance in the textile industry In order to assess the feasibility and efficacy of alternative interventions to promote a shift to sustainable industrial development, it is useful to examine the nature of the governance system in the textile industry. This will provide information on the mechanisms and processes that influence the allocative decisions of a single entrepreneur.

The governmental machinery pertaining to the industrial sector is neither as sophisticated nor as complex as that for the agricultural sector. First, while there is a national system of industrial research institutions, the Pakistan Council for Scientific and Industrial Research (PCSIR), it is neither as goal-driven nor as well integrated into the production system as its counterpart, the Pakistan Agricultural Research Council. Second, there is no integrated system for determining industrial policy, especially when it comes to environmental considerations. Third, the system of industrial education (both engineering and business schools) are more or less completely divorced from the research and policy institutions. Finally, although the environmental agencies have tended to focus on the industrial sector, their clout and institutional strength are far from adequate given the size of the task.

Having said all this, the very weakness of the policy and research establishment in the industrial sector has made it possible for inroads to be made in designing environmental policy. In the agricultural sector, the research and policy establishment is at this moment an obstacle to proper policy development. However, the existence of an organized industrial association made it possible for such an initiative to have legitimacy and influence. This can enable us to distinguish between three different types of governing structure in the textiles sector.

For this purpose, it is useful to recall the distinction between various segments of the industry. As mentioned earlier, the yarn-spinning industry is large-scale and well organized, and is represented nationally by a strong lobbying group, the APTMA. These units sell their products to importers or to local exporters. There is not much evidence of a consumer-driven commodity chain in this segment. The relationship between buyer and seller is typically arm's length and transitory. The most direct route to influencing the spinning sector is the APTMA.

Weaving is an 'anarchic' sector, dominated by small-scale, informally organized firms. They do not have a strong lobbying presence at the national level, nor are their links with cloth buyers of a longer-term nature. These two sectors are not of major environmental concern.

The area of most concern is the textile-processing sector, which has effluent treatment obligations under the new laws. It is also subject to restraint because of the ban on azo dyes. This is a relatively small sector, consisting of large, professionally organized firms. The experience of the roundtable suggests that these units are susceptible to influence by government policy. In this case, the relative lack of influence by the organized association and the absence of a coherent governmental policy-making structure can allow a governing system to emerge. This system would be a coalition of various stakeholders in the process. As evidenced already, such a collective system has the potential of influencing production behaviour.

Given that all these segments are tied into the global market, another potential source of influence is the Ministry of Commerce and the Export Promotion Bureau. However, so far these agencies have played a servicing rather than a leading role in the industry.

The segment that is closely tied into the global commodity chain is that of garment and apparel production. In this segment, market conditions are determined by a small number of international retail chains. In recent years, these retailers have developed mechanisms to influence the in-plant allocative decisions of even small firms. Much of the recent concern has focused on labour standards issues. All major retailers have arrangements through which to evaluate compliance with international labour standards within their suppliers. The assessment is carried out through scheduled visits by designated buyers from the region (often East or Southeast Asia). Large retailers generally do not schedule repeat visits by their buyers for a firm that fails inspection. The object of the assessment is to examine occupational health and safety considerations, child labour requirements and compliance with other labour standards.

The successful record of garment buyers in establishing working relationships with foreign vendors is in marked contrast to the experience of similar industries: surgical goods and carpets. In both these cases, the local market structure is very similar. Small-scale, informally organized firms, without adequate national collective organizations, dominate the industry. In both, there have been problems stemming from the violation of health and safety requirements, or child labour legislation. However, in neither case has there been a successful resolution.

This brings us back to Gereffi's (1994) idea that industrial concentration is linked with the emergence of governance system in commodity chains. Given the presence of such a market structure, the large Northern retailers have the capacity to influence production processes in an effective manner. However, their reach extends only to garment suppliers, and there is very little interaction with other units.

Prospects of a transition The main issue, therefore, is the mechanism through which the system of production can be transformed in order to become more sustainable. One mechanism is through consumer influence, as exerted in the form of price differentials, boycotts or even trade restrictions to reflect consumer preference. Another mechanism is the use of technical assistance, research, extension and specialized inputs, as in the Green Revolution. Clearly, where it is possible for an indigenous agency to play the role of underwriting the system of governance, the second option might be both feasible and efficacious. Where the system of governance is led by large retailers, who can set up their own technical assistance and information clearing arrangements, the transformation can be brought about by the threat of consumer action.

In this sense, the government and the retailers are quite alike, since both require the existence of the threat of sanctions to be effective. The government uses its access to national resources, increasingly limited as they might be, and the threat of legal action. Corporations use their access to markets, and the threat of exit, to discipline suppliers. The systems work well only where inspection systems and information flows are relatively efficient.

The prospects of transition to sustainable industrial growth depend crucially on the nature of governance in the specific segments of the industry. The segments that are well organized, spinning and processing, can take advantage of emerging opportunities and protect themselves from adverse changes. However, the history of their interaction has been one of pressuring the government for protection rather than adapting to changing market conditions. Still, there is considerable scope for action by the government to influence their behaviour. The SDPI initiative has chalked out a mechanism through which such intervention can become legitimate as well as effective.

In the case of the garments industry, changes can indeed be brought about directly. This can be most effective where the issue is one of incremental rather than systemic change, as for example in the case of the azo dyes. Given that switching from one kind of dye to another does not require a transformation of the production system, it is likely that the transition will take place smoothly and efficiently to comply with the demands of the international importers. However, changes of a systemic nature require broad-based technical assistance to upgrade management and production arrangements.

The absence of strong backward institutional linkages means that the ability of consumer-driven chains to drive the entire system is fairly limited. In this case, it might be more appropriate to design an alternative mechanism that borrows from the Green Revolution model of technical change. Instead of a pre-established network of research and extension dominated by the government, this would rely on improved communication between stakeholders, and the establishment of links between existing institutions of research and extension.

Notes

1. Senior research director, Stockholm Environment Institute/Tellus, Boston. Chapter prepared for UNEP project on trade and environment. Tahir Hasnain and Shahid Zia provided invaluable assistance and advice in the preparation of this chapter. Comments and advice are also gratefully acknowledged from the coordinators of the UNEP project (Konrad von Moltke, Onno Kuik, Nicolien van den Grijp) on commodity chains, from participants in two project workshops at Islamabad (September 1997) and San José (January 1998) and from numerous experts in Pakistan.

2. Southern critics do not question the negative environmental impact of cotton production, especially for producing countries. They question, rather, the selective focus on Southern industries, and lesser criticism of equally or higher polluting Northern practices (e.g. energy-intensive transportation

systems). Others point out that cotton is more environmentally benign than its synthetic alternatives. For example, in a study prepared by the sports clothing firm Patagonia Inc. cited in IISD/WWF (1997: 76), cotton had the lowest rating of environmental risk severity (3.45), followed by wool (4.36), and nylon and polyester (4.73 each). In citing this study, the IISD/WWF warn of a possible conflict of interest given that Patagonia is a clothing manufacturer. However, probably a higher proportion of its revenues derives from clothing made from synthetic products.

3. Gereffi (1994: 99) notes the relationship of his analysis with the concept of 'flexible specialisation' (see Piore and Sabel 1984). Another related concept is that of 'collective efficiency' introduced by Schmitz (1989) and his colleagues at the University of Sussex. They see the informal sector organized in such a manner as to enhance the collective efficiency and profitability of a community of producers. This is very similar to Gereffi's analysis, even though it suggests a decentralized form of governance.

4. Marglin is careful to say that these two ideal types are present in all forms of knowledge production. However, the modern world view sees *episteme* as the only legitimate form of knowledge and *techne*, at best, as practice waiting to be transformed into true knowledge.

5. For a good description of cotton production, see IISD/WWF 1997.

6. According to an expert interviewed by us, large retailers have inspection agents in various parts of the world. Agents located in Singapore or Hong Kong often cover Pakistan. These agents schedule inspection visits, and if the firm is rejected even for a small violation, it is almost impossible to get the company to schedule another visit.

7. The term 'pesticides' includes insecticides, nematicides, fungicides, herbicides, defoliants and desiccants, but not fertilizers. In Southern countries, the most significant reliance is on insecticides.

8. The entire range of activities includes residue disposal, control of pests, return of nutrients to the soil, pre-plant tillage, seedbed preparation (including fertilizer and herbicide application), planting (along with application of fungicides, herbicides, pesticides and fertilizer), weed and insect control (including defoliation if harvesting is by mechanical means), harvesting (by manual or mechanical means), transportation, ginning (i.e. removing seeds from lint) and baling.

9. The entire sequence of activities covers spinning (blowing, mixing, carding, combing, drawing, simplex, ring spinning and cone winding), weaving (warping, sizing and weaving), processing (singeing, desizing, scouring, mercerizing, bleaching, dyeing, printing and finishing) and garment manufacturing.

10. This is consistent with broader evidence that shows that textile weaving and processing are undertaken by large integrated factories, while garment manufacturing is fragmented and small scale in structure (see Gereffi 1994: 101–2).

11. This assumes, however, that the reduction in total output will not raise

final prices. In fact, the aggregate impact of the change on final prices is more complex and beyond the scope of this exercise.

12. These are very rough calculations, and should be treated with caution. They do not represent actual shares of domestic consumption and external trade in various segments of the cotton chain.

13. The vast majority of the longer staples are produced in only five countries: the USA, Peru, Egypt, Sudan and the former USSR. These varieties have higher water requirement and a longer growing period, the latter making them more susceptible to insects and diseases (de Vries 1995).

14. Two-thirds of the weight of the cotton plant consists of cottonseeds, which contain valuable nutrients (18–25 per cent fat and 29–34 per cent protein), and are used as cattlefeed or in the production of cooking oil. Average to moderate cotton yields of 1,500 kilograms per hectare of 'seed cotton' (i.e. including both seeds and lint) produce not only 500 kilograms of lint, but also 1,000 kilograms of cottonseeds, which contain the same caloric content as a normal harvest (in Southern countries) of 600 kilograms of cereal from the same area. However, this creates some confusion over yield figures, which can refer to lint cotton (without seeds) or seed cotton (lint plus seeds). A yield of 1,500 kilograms per hectare is high if it refers to lint cotton, but moderate if it refers to seed cotton. Official documents typically refer to lint cotton yields, but some influential writings (e.g. Murray 1994) have used seed cotton yields without saying so explicitly.

15. For a discussion of the role of various chemical inputs in cotton production, see IISD/WWF 1997: 11–12). Murray (1994) provides an excellent description of the growing dependence on pesticides in Latin America, and the health and other problems caused by them. This and the following two paragraphs are drawn mainly from Murray 1994: 12–16.

16. This is quite typical for Southern countries, where cotton often uses up to 85 per cent of all pesticides. Global figures are variously estimated at between 11 and 25 per cent.

17. Presumably, the area figures represent multiple spraying, since the area covered by commodities that use pesticides (cotton, fruits and vegetables) is far less than the total reported area. Cotton is grown on roughly 3 million hectares. The total area allocated to fruits and vegetables was less than 900,000 hectares between 1991 and 1996.

18. For example, heavy use of DDT, methyl parathion and toxaphene to combat the American bollworm in El Salvador contributed to the outbreak of the whitefly. The poor contact of the sprayed insecticides with the nymphs and pupae of the whitefly (which are on the underside of the leaf) is a reason for its resurgence.

19. PL–480 is the local provision of counterpart funds generated by the US supply of food aid.

20. As evidenced by the fact that pest control involved only 2.5 per cent of the variable costs (excluding land costs) under conventional methods. Normal

levels of pest control costs range between 7 and 57 per cent of variable costs. See de Vries 1995: 55–6.

21. For example, if the value of the crop depends on its perceived quality, the decisions of others will affect the revenue received by a single farmer. These types of externalities are determined by such factors as reputation, homogeneity, costs of transportation and location of markets.

22. A large number of farmers purchasing a single input will make it profitable for suppliers to supply it in bulk and presumably at a lower cost.

23. For an interesting anthropological analysis of the changing nature of agronomic advice in Western India, see Appadurai 1990. He notes that while 'agriculture' is a form of 'culture', namely a way of thinking that integrates values and behaviour, the subject of the programme was 'agronomy', namely technological knowledge of the agricultural system, independent of the cultural context.

24. The institutions did not cover all significant aspects of the production process, however, and the current agricultural crisis is associated directly with the areas where the institutions either did not exist or remained outside the self-reinforcing circle. These areas are water (which remained with a completely independent department of irrigation), social, economic and equity issues (for which there are specialized institutions within the PARC system, but neither well integrated nor well regarded by the research establishment), and environment (which was not given any importance).

25. Especially the emergence of strong political movements, democratic and egalitarian rhetoric of nationalist political parties, expanding education and information, population growth and the strengthening of the mass media and other organs of civil society generally.

26. This section is based on SDPI 1995.

27. For example, over 50 per cent of the credit from government-controlled development finance institutions (DFIs) goes to spinning units. Similarly, higher duties on inputs of processing and finishing units render them less attractive than spinning units.

28. Pakistani yarn exports registered an increase after Pakistan joined the WTO in April 1994, and benefited from an enhancement in the yarn quota to the European continent from 9,000 to 15,000 tonnes.

29. Share prices reached rock bottom after Pakistan's nuclear tests led to the freezing of foreign currency accounts, and then started rising fairly rapidly after the military take-over of 12 October 1999.

Bibliography

Appadurai, A. (1990), 'Disjuncture and difference in the global cultural economy', *Public Culture*, Vol. 2: 1–24.

APTMA (1996), APCOM Series No.150, Lahore, Pakistan.

Bashir, A., M. A. Chaudhary and S. Hassan (1994), *Cost of Producing Major Crops in the Punjab*, Faisalabad: University of Agriculture.

Chaudhary, A. M., B. Ahmad and M. A. Chaudhary (1992), *Cost of Producing Major Crops in Pakistan 1991–92*, Faisalabad: University of Agriculture.

CUTS (1997), *Textiles and Clothing: Who Gains, Who Loses, and Why?*, briefing paper, Calcutta: CUTS Centre for International Trade, Economics and Environment (CITEE).

Daniel, Alexander, Bo van Elzakker and Tadeu Caldas (1999), 'India', in Dorothy Myers and Sue Stolton (eds), *Organic Cotton: From Field to Final Product*, London: ITDG Publishing.

Dinham, B. (1993), *The Pesticide Hazard: A Global Health and Environmental Audit*, London: Pesticides Trust.

Edwards, A. (1993), 'The impact of pesticides on the environment', in D. Pimental and H. Lehman (eds), *The Pesticide Question: Environment, Economics, and Ethics*, New York: Chapman and Hall.

Eisa, H. M., S. Barghouti, F. Gillham and M. Tawhid Al-Saffy (1994), *Cotton Production Prospects for the Decade to 2005: A Global Overview*, World Bank Technical Paper No. 31, Washington, DC: World Bank.

Gereffi, G. (1994), 'The organisation of buyer driven global commodity chains: how US retailers shape overseas production networks', in G. Gereffi and M. Korzeniewicz (eds), *Commodity Chains and Global Capitalism*, Westport, CT: Praeger.

Government of Pakistan (1983), *A Study on Cost of Production of Crops: Cotton (1980–81)*, Islamabad: Ministry of Food, Agriculture and Cooperatives.

— (1995), *Agricultural Statistics of Pakistan 1993–94*, Islamabad: Ministry of Food, Agriculture and Livestock, Economic Wing.

— (1996), *Support Price Policy for Seed Cotton, 1996–97 Crop*, Islamabad: Agricultural Prices Commission.

— (1997a), *Economic Survey 1996–97*, Islamabad: Ministry of Finance.

— (1997b), *Fifty Years of Pakistan in Statistics*, Islamabad: Federal Bureau of Statistics.

Hussain, T. and M. Ali (1975), 'A review of cotton diseases of Pakistan', *Pakistan Cotton*, Vol. 19: 71–86.

Hussain, T. and T. Mahmood (1988), 'A note on leaf curl disease of cotton', *Pakistan Cotton*, Vol. 32: 248–51.

IISD/WWF (1997), 'The cotton industry: towards an environmentally sustainable commodity chain', report prepared for the Workshop on Cross-National Environmental Problem-Solving, School of International and Public Affairs, Columbia University.

Ingco, M. D. and L. A. Winters (1995), 'Pakistan and the Uruguay Round: impact and opportunities: a quantitative assessment', International Economics Department, Trade Division, Background Paper for Pakistan 2010 Report, Washington, DC: World Bank.

Jabbar, A., S. Z. Masud, Z. Parveen and M. Ali (1993), 'Pesticide residues in cropland soils and shallow groundwater in Punjab, Pakistan', *Bulletin of Environmental Contamination and Toxicology*, Vol. 51: 268–73.

Kishor, N. M. (1992), 'Pesticide externalities, comparative advantage, and commodity trade – cotton in Andhra Pradesh, India', Washington, DC: World Bank, Country Economics Department.

Korzeniewicz, R. P. and W. Martin (1994), 'The global distribution of commodity chains', in G. Gereffi and M. Korzeniewicz (eds), *Commodity Chains and Global Capitalism*, Westport, CT: Praeger.

Low, P. (1995), 'Pakistan: the Uruguay Round and trade policy reform into the next century', Background Paper for Pakistan 2010, Washington, DC: World Bank.

Marglin, S. A. (1990), 'Losing touch', in F. Apffel Marglin and Stephen A. Marglin (eds), *Dominating Knowledge,* Oxford: Clarendon Press.

Murray, D. L. (1994), *Cultivating Crisis: The Human Cost of Pesticides in Latin America*, Austin: University of Texas Press.

Network, The (1995), 'From Cotton to Textiles to Clothing', *The Network*, Vol. 8: pp. 1–26.

Pakistan Central Cotton Committee (1997), *Cotton Review*, Karachi: Ministry of Food and Agriculture, Government of Pakistan.

PARC (Pakistan Agricultural Research Council) (1992), *Annual Report of the Pakistan Agricultural Research Council 1992–93*, Islamabad: PARC.

Patagonia Inc. (1991), *Developing Corporate Environmental Management Program*, Werner International and RCG/Hagler Bailly, Inc.

Piore, M. and C. B. Sabel (1984), *The Second Industrial Divide: Possibilities for Prosperity*, New York: Basic Books.

Schmitz, H. (1989), 'Flexible specialization: a new paradigm of small scale specialization', IDS Discussion Paper, No. 261, Brighton: Institute of Development Studies.

SDPI (1993), *Environmental Examination of the Textile Industry in Pakistan, Project on Technology Transfer for Sustainable Industrial Development*, Islamabad: SDPI.

— (1997), *Cotton File 1995–96*, Islamabad: SDPI.

de Vries, H. (1995), 'An international commodity related environmental agreement for cotton: an appraisal', Amsterdam: Vrije Universiteit, ICREA Research Team.

Zeegers, J. (1993), 'Duurzame teelttechnieken in de koeten', master's thesis, Vrije Universiteit, Amsterdam.

CHAPTER 4

Environmental Impacts and Mitigation Costs: The Case of Pakistan's Cloth and Leather Exports[1]

Shahrukh Rafi Khan, Mahmood A. Khwaja, Abdul Matin Khan, Haider Ghani and Sajid Kazmi[2]

§ Pakistan's commitment to environment and sustainable development is outlined in its National Conservation Strategy (1992). The authors of this, not unlike the authors of the World Conservation Strategy, could not foresee the pervasive impact of trade on the environment. Indeed, the Ministry of Commerce was not represented on the steering committee. The representation from NGOs and the private sector did not reflect this aspect either. Nor was there any effort to commission a background paper by outside experts.

Thus, for trade and environment, the situation was one of two distinct cultures. Knowledge and postures existed separately, with a conspicuous lack of a cross-cultural view. Yet the need, in the wake of the WTO work programme, is for a cross-cultural view. Before this can be accomplished, it is important to expose policy-makers, NGOs and the private sector to the main issues involved in the debate on trade and the environment and the findings of primary research on key areas in this field outlined in its *National Conservation Strategy*.

There also appears to be a misperception on the part of political authorities in Pakistan that cleaning up the environment is a luxury we cannot afford or that preventing environmental damage imposes an unbearable economic cost. This is true only when viewed from a limited short-term perspective. Politicians and businesses need to realize that environmental damage depletes the natural resource base via water, soil and air degradation and results in current and future loss of productivity. Much more important is the loss of productivity resulting from the impairment of the health of current and future generations.[3] Politicians always speak for the poor, but it is the poor who are least capable of defending themselves against environmental ravages. If improving the health, productivity and quality of life of the current and future generations is not a sufficient inducement to act quickly, the potential huge loss of export markets should be. The Uruguay Round-induced increase in exports for developing and transitional economies has been estimated to be between 14 per cent and 37 per cent.[4] Thus the dividends from the right decisions are potentially very high.

An analytical framework developed by the OECD (1994: 7–17) categorizes the environmental impacts of trade into product, scale, structural and regulatory effects. In each category, there can be positive and negative effects. Our focus is on the negative-scale effects that can result from trade expansion and trade liberalization in two of Pakistan's key manufacturing export sectors. Thus, as production expands to respond to growing export markets, without proper environmental policy and enforcement mechanisms in place, these enhanced exports will prove to be environmentally disastrous. Fortunately, in Pakistan's case, a reasonable environmental policy is in place. Currently, government, business and civil society groups are groping towards appropriate implementation mechanisms. This research will indicate the urgency of timely implementation. We will also demonstrate the cost and benefits of mitigation strategies. Our main finding is that the costs of mitigation are much lower than perceived to be the case in the South, and also much lower than the social benefits.

Research Method

The overall objective is to do a heuristic cost–benefit analysis of the abatement of the incremental pollution resulting from cloth and the leather industry exports.[5] The following four-step procedure has been adopted:

1. Estimate the increased cloth and leather exports up to the end of 2004, when textile and clothing quotas in developed countries are expected to be removed as negotiated in the Uruguay Round Agreement on Textiles and Clothing (ATC). While, in principle, this represents an important date for our research, its significance is somewhat reduced since 72 per cent of Pakistan's cloth exports go to non-quota countries.[6]
2. Estimate the environmental impact of cloth, leather and footwear exports. By using unit discharge rates of chemical, organic and

suspended pollution loads, based on data collected by the Sustainable Development Policy Institute's (SDPI) Technology Transfer for Sustainable Industrial Development project (TTSID) and the Environment Technology Programme for Industry (ETPI) predict the effluent pollution associated with exports. It would have been useful also to assess the total production-related pollution and mitigation cost. However, recent production data in Pakistan are not available, since the last Census of Manufacturing Industries took place in 1990. The textile and leather plants were purposively selected and can be viewed as roughly representative of medium units in Pakistan.[7]

3. Assess the import costs of using cleaner technologies. The technologies referred to for the textile sector are the ones best suited to local conditions to meet the currently applicable environmental quality standards in Pakistan.[8] The technology considered for the leather sector is locally available.

4. Assess the mitigation impact of using cleaner technologies and set that in an understandable context for business and government.

We also document the health and other social costs resulting from pollution. In effect, the reduction of such costs represents the benefit from pollution mitigation. While it is not possible specifically to link the health costs to incremental export-related leather and cloth production, research has been conducted on quantifying the cost of pollution on an aggregate level. Based on the shares of leather and cloth production in total value added, and breaking that down further by exports as a per centage of total cloth and leather production, we attribute pollution costs to exports by these industries and hence the implicit benefit from mitigating the pollution.

Justification for Industry Selection

The rankings in Table 4.1 show that textiles are clearly the sector of major economic importance to Pakistan in all categories. While leather is not quantitatively of similar significance, it clearly is so from an environmental perspective, as the following section indicates.

Environmental Impacts

Environmental impacts of cotton exports[9]   In investigating the environmental impacts we start with cotton production, which is where the commodity chain begins.[10] Two of the most damaging inputs into cotton production are pesticides and fertilizers, and so we start the analysis with the environmental impacts of these inputs.

PESTICIDES The main negative environmental impact from cotton production results from the use of chemical inputs. Rachel Carson's *Silent Spring* (1962) started the questioning and many writers have since written about the negative effects of pesticides,[11] particularly concerning their use in developing countries.[12] Weir and Schapiro in *Circle of Poison* (1981: 11) pointed out that pesticide poisoning in LDCs was 13 times greater than in the USA, due to the lower level of education, despite much greater use in the USA. Drifting pesticide sprays, leaky applicators, inappropriate use and over-use result in run-offs and seepage into water and soil.

Residues in soil, food and water and unsafe handling result in various medical problems for people, including enzyme imbalances, skin complaints, allergic reactions, delayed neurotoxicity, behavioural changes, lesions, changes in the central nervous system, peripheral neuritis, cancer, sterility, cataracts, lung perforations, memory loss and damage to the immune system. Colburn (1994: 89) stated that most of the past testing focused on individuals directly exposed and not on the functionality of their offspring. He states that studies reveal that 'as a result of [pesticide] exposure in the womb of mammals including human, the endocrine, immune and nervous systems of embryos do not develop normally'.

Sadhu (1992: 23) cites an FAO study claiming that only 5 per cent of the insecticide falls on target plants; the rest pollutes the environment.[13]

Table 4.1 The economic significance of the textile and leather industries in Pakistan

	Textiles and clothing	Leather and products
Exports as % total exports[1]	55.0 (1)	3.0 (4)
Value added as % of total value added in major industries	27.7 (1)	1.6 (15)
Employment as % of average daily employment in major industries	41.5 (1)	2.4 (8)

Notes: Parentheses contain ranks. 1. Pakistan's share of world exports of yarn and cloth in 1995 were 28.3 and 5.8 per cent respectively according to the *Cotton World Statistics*, quarterly bulletin of the International Cotton Advisory Committee, Vols 35, 45 and 48.

Source: Government of Pakistan 1997: 74–5. Export figures are taken from Government of Pakistan 1996: 29–30, 338.

The adverse effects on the land base include a reduction in the natural fertility of the soil, harm to the soil structure and soil aeration, reduction of the water-holding capacity of the soil, making it more prone to erosion by water and wind, and lower drought tolerance of crops. Finally, pesticides are viewed as indiscriminately killing useful insects, micro-organisms and insect predator species, breeding more virulent and resistant species of insects and vectors and reducing the genetic diversity of plant species.[14]

In Pakistan, there is evidence that cotton pests such as the American bollworm and the whitefly have developed resistance against common pesticides, and this had a devastating economic impact in Pakistan's mono-economy in 1992–93. Sale of adulterated pesticides is perceived as one cause of such resistance.[15] This kind of phenomenon results in what has been referred to as the 'pesticide treadmill' whereby farmers feel compelled to use more pesticides when less does not work and more is perceived to be better if less is working (see Chapter 3). In addition, aggressive marketing by multinational pharmaceutical companies leads to over-use and also to a market for adulterated pesticides sold at lower prices.

Jabbar and Mallick (1994) review the scanty evidence on this issue in Pakistan and report the existence of residues in water, soil, food and people.[16] This evidence also indicates the existence of the above-mentioned maladies resulting from pesticides.

FERTILIZERS   Qutub (1994: 16) documents the costs to human health and the environment. Excess nitrate and nitrite in water and foods can result in methemoglobinaemia ('blue baby' syndrome) in infants, are viewed as carcinogenic and can result in respiratory illnesses. Run-off can result in eutrophication via enhanced algae growth and hence hurt fish stocks and also humans via algae toxins. Soil erosion can result from volatilization and denitrification. Finally, nitrates contribute to soil-pan formation and nitrogenous gases can contribute to the greenhouse effect.

Fertilizer use in Pakistan has steadily increased from 20 kg per hectare in 1971–72 to 91 kg per hectare in 1991–92 and 103 kg in 1994–95. Evidence on the negative environmental impact of fertilizers in Pakistan is once again very limited. Ali and Jabbar (1992: 92) tested soils in Faisalabad in a pilot study and concluded that nitrates are present in sub-surface soils in considerable quantities.

ANTICIPATED INCREASE IN INSECTICIDE AND FERTILIZER USE
Since farmers do not use herbicides or defoliants, the main source of

concern is the use of insecticides.[17] The consumption of pesticides in 1997 was 44,872 metric tons[18] and a large portion is used in cotton production (about 65 per cent).[19] Pakistani farmers use 8–13 sprays per season, which is about twice the level recommended by cotton researchers and extension staff. The number of sprays and the area covered has increased dramatically over time. Thus while the area sprayed as a percentage of total area under cotton cultivation was 5–10 per cent in 1983, it was 95–98 per cent in 1991.

To get a handle on the quantitative increases in fertilizer and pesticide use associated with cotton production, we adopted the following approach. Much of the cotton produced gets exported, either directly as raw cotton or indirectly as cotton products. Also, almost all the pesticides are imported. Thus the changes in chemical input use can broadly be viewed as being trade-related.

$$CRIU = \phi \delta IU$$

where

CRIU = forecast of growth in cotton-related chemical input use

ϕ = cotton production share in total chemical input use

δIU = forecast of growth in chemical input use

Cotton production's share in pesticide and fertilizer use in the base year (1996–97) was about 65 per cent and 50 per cent.[20] We assume that this continues to be cotton production's total share in pesticide use in the terminal year. Having an estimate of ϕ, one can simply multiply that with the increase in chemical input use (δIU) to get an estimate of the change in chemical input use that can be attributed to an increase in cotton production.

We used the auto-regressive, integrated, moving average approach

Table 4.2 Forecast increase in chemical input use due to increase in cotton production

Input	Base year (1997)	Terminal year (2004)	Change in input use (1997–2004)	Change in input use attributed to cotton production
Pesticide (MT)	44,872	63,192	18,320 (40.8)	(31.7)
Fertilizer (NT)	2,409,000	3,480,000	1,071,000 (44.4)	(22.2)

Notes: MT = metric tonnes; NT = nutrient tonnes; parentheses contain growth rates.

Source: Table 4.3 (below).

(ARIMA) to generate the forecasts of the right-hand side of the equation above, i.e. of the increase in cotton production to get ϕ and of the increase in chemical input use to get δIU. In Table 4.2, we present the base and terminal year pesticide and fertilizer consumption and the expected contribution of cotton production to the increase in chemical input use.

Table 4.3 shows how using the above method we arrived at the ARIMA forecasting model and estimate for the variables included in our analysis.

Projecting from past trends, pesticide and chemical fertilizer use is expected to continue to increase in Pakistan. Chemical fertilizer use is much more intensive in Japan and Europe, with the Netherlands applying the most (554 kg per hectare) in 1994–95 compared to Pakistan's 103 kg per hectare. However, while use in the major OECD countries is much higher, use in all of them has been steadily declining since the middle to late 1980s and use in the USA is already as low as that in Pakistan.[21] Pakistan does not need to wait for the same intensity of use to derive the same lessons, because well-known alternatives such as the integrated plant nutrient system (IPNS) and integrated pest management (IPM) are already available (see Chapter 3).

Environmental impacts

COTTON AND TEXTILE Industrial pollution can be categorized either according to the medium through which it enters the environment –

Table 4.3 ARIMA model estimate for the forecast for 2004

Variable	ARIMA model[1]
Pesticide consumption (metric tonnes)	(0,1,1)
Fertilizer consumption ('000 N/tonnes)	(0,1,1)
Cloth (million sq. metres)	(0, 0, 0) OLS
Hides and skins ('000 kg)	(0,0,1)
Leather (million sq. metres)	(0,1,2)
Footwear (million pairs)	(0,1,2)

Note: 1. Represents the ARIMA model (p, d, q) selected, where p represents the number of autoregressive terms, d the number of times a series has to be differenced to make it stationary and q the moving average terms.

Sources: For pesticide consumption, Government of Pakistan 1998a: 155; for fertiliser consumption, Government of Pakistan 1998b: 59. The latter source (pp. 168–70) was also utilized for the remaining variables.

air, water and soil – or according to the processing stage. Of the three major industrial processes in textile processing – spinning, weaving and finishing – environmental problems are associated mainly with the last stage. Spinning entails mostly dry processing and virtually no harmful effects are generated. In the weaving process, starch is applied to the fabric to impart strength and stiffness, resulting in wastewaters that contain large amounts of starch with high BOD values. The finishing stage uses a variety of chemicals, including acids, alkalis, wetting agents and chemical dyes. These effluents require proper treatment before being discharged into an external drain.

Air pollution and airborne wastes are not a major problem in textile production. The effects of air emissions are fairly limited and localized, although if they are not properly managed, they can be harmful to the health of workers. These can be classified into four categories: oil and mists, dust and lint, solvent vapours, and odours. Dust and lint problems are peculiar to the spinning stage. However, the yarn-spinning sector in Pakistan is generally equipped with self-contained waste recovery units. These units reduce particulate emissions and health risks to workers, and improve the working environment as well as the net profitability of the enterprise. By virtue of these units, lint losses in the spinning section are virtually nil. A typical spinning firm processes 60 tonnes of yarn per day, out of which 11 per cent is recoverable as saleable yarn waste, and another 9 per cent recovered and converted to low-grade yarn. Similarly, these units maintain temperature, humidity and noise levels well within safe limits. However, small textile units in the un-organized sector do face a serious air pollution problem, due to the lack of modern plant and machinery. Workers at these units remain exposed to contaminated air, and thus to the risk of lung disease and asthma.

Textile processing includes a number of wet processes – bleaching, mercerizing, dyeing and finishing – that produce liquid effluent with varying waste composition. Environmental quality standards are most developed in the case of liquid effluent. At present, the majority of the textile mills, including modern, integrated facilities, do not have ad-equate arrangements to treat their effluent before discharging into an external drain. Since, in many cases, the external sink is an irrigation canal, the untreated chemicals can affect the quality of irrigation water. Also, textile processing is a heavy user of water.

An average integrated textile mill produces 15 tons of finished cloth per day. It uses a total of approximately 3,840 cubic metres of water per day, including 1,680 cubic metres for finishing and processing, another 960 cubic metres for steam generation, and an equivalent volume for serving the workers' colony and other domestic uses of water. The water used for finishing and processing results in contaminated liquid effluent of approximately 1,500 cubic metres per day. In addition, the water used for domestic use is emitted with household sanitary wastes.

Environmental quality standards are classified into aggregate measures and maximum allowable concentrations of specific chemicals.[22] The aggregate measures are the pH value (which determines the acidity or alkalinity of the effluent), temperature, the biological oxygen demand (BOD),[23] the chemical oxygen demand (COD),[24] the total suspended solids (TSS) or non-filterable residue, total dissolved solids (TDS)[25] and colour.

Estimates of characteristics of industrial effluent and the cost of chemical treatment to bring them within allowable limits have been carried out in Pakistan in a small number of projects. The most notable among these is an innovative project led by the Sustainable Development Policy Institute (SDPI). The project, entitled Technology Transfer for Sustainable Industrial Development (TTSID), was funded by the Swiss Federal Office for Foreign Economic Affairs. Some results are provided in Table 4.4.

The average Pakistani mill emits a grey liquid effluent with high pH value (8–10), BOD (112–120), COD (430–480), TSS (25–1,200), TDS (2,300–3,600), and temperature (52° Celsius). These are generally outside the range of environmental standards in Pakistan, and also in compar-

Table 4.4 Textile industry pollution levels and environmental standards

Parameters	Measured level mg/l	Pakistan NEQS mg/l	World Bank guidelines mg/l	Other standards Indian mg/l	USEPA mg/l
PH	8–9	6–10	6–9	5.5–9	6–9
BOD	112–120	80	58	150	58
COD	430–480	150	524	–	524
TSS	26–1,200	150	157	100	157
TDS	2,300–3,600	3,500	–	–	–
Chromium	0.05–0.30	1	0.90	2	0.9
Phenol	not detected	0.1	0.90	1	0.9
Sulphides	0.07–15.0	1	1.75	2	1.75
Temp. °C	52	40	5+amb.	40	5+amb.

Source: SDPI 1995: 25.

able countries. The effluent also contains variable amount of sulphides and chromium.

LEATHER TANNING   The main source utilized for the first three paragraphs of this sub-section is ETPI 1997, which drew its information from an audit of three tanneries. Parikh et al. (1995) was also extensively drawn on. Leather tanning has been ranked as one of the most polluting activities compared to other manufacturing sector activities. It also has one of the highest toxic intensities per unit of output.[26]

Converting hides into leather is a heavily chemical-intensive process utilizing roughly 130 chemicals. The main chemicals used in the various processing stages include sodium sulphide, lime powder, ammonium sulphate, sodium chloride, sulphuric acid, chromium sulphate, sulphonated and sulphated oils, formaldehyde, pigments, dyes and anti-fungus agents. The processing stages are pre-tanning (soaking, de-hairing and liming, fleshing, de-liming, washing, bating and de-greasing), tanning (pickling, chrome tanning, wet-blue storage, sorting, splitting and shaving), wet finishing (wet back, neutralization, re-tanning, washing, fat liquoring, dyeing and washing), dry machine process (sammying/ setting, drying, stacking/toggling, shaving, trimming and pressing), and finishing (buffing, spraying/coating, drying and glazing/polishing).

Pollution or wastes resulting from these processes are airborne solid and (primarily) liquid. Hydrogen sulphide and ammonia are the major gases released into the atmosphere. However, laboratory results showed emissions lower than the national environmental quality standards.

Most of the solid wastes are recycled. The drums, cartons and chemical bags are re-used. Fleshing, raw trimming and buffing dust is bought by leather board or poultry feed manufacturers. These solid wastes contain chromium residue, which is known to cause lung perforations and bronchiogenic carcinoma in humans who are continuously exposed. Chicken feeds prepared from proteins containing tanneries' solid wastes is likely to cause direct entry of chromium into the food chain. The results of tests conducted by the Pakistan Tanners Association showed chrome residues in poultry feed. Leather shavings are used as cheap fuel in kilns, causing the release of chromium into the environment. The remaining solid wastes are usually illegally dumped around the factory area on unutilized lands. These solid wastes include metal contents such as chromium, aluminium and zirconium, which have a detrimental effect on plant growth.

In the course of processing of hides into leather, roughly 50–150 litres of water are used per one kilogram of converted leather. Thus effluents discharged from tanneries are voluminous and highly coloured, and contain a heavy sediment load including toxic metallic compounds, chemicals, biologically oxidizable materials and large quantities of putrefying suspended matter. Tannery effluents, without any pre-treatment, are discharged indiscriminately into water bodies or open land, resulting in contamination of surface as well as sub-surface water. The lack of effective implementation of legislative control, poor processing practices and the use of unrefined conventional leather-processing methods have further aggravated the pollution problem caused by the tanning industry in the South Asian region, including Pakistan.[27]

As in the case of textile effluents, the low pH of tannery effluents causes pollution of the water system. Large pH fluctuations and the high BOD value caused by tannery effluents can kill all natural life in an affected water body. Studies have revealed that the water of the River Ganges at Kanpur and the sub-surface water of the Paler river basin of India and Korangi and the Charsadda regions of Pakistan have been significantly polluted by tannery wastes.[28] The contribution of tannery pollution to the contamination of the Karachi Coast is estimated at about 10–15 per cent of the total pollution. Hydrogen sulphide (formed due to the presence of sulphide in the effluent) and chromium are highly toxic to many life forms. Some workers died in Karachi in 1980 while clearing monsoon ditches filled with tannery sludge.[29]

In the Pakistani Punjab and the Palar river basin in India, tanneries are directly contaminating prime agricultural land. Research has shown that the crop yield has been adversely affected and also, of course, the food is contaminated.[30] Most of the tanneries in Punjab and NWFP in

Table 4.5 Measured contamination levels and discharge standards in the leather sector

Parameter	Measured level mg/1	Pakistan EPA standard mg/1	USA EPA standard mg/1
pH	7.4–9.8	6–10	6–9
BOD	1,740–11,050	80	58
COD	3,800–41,300	150	524
TSS	440–890	150	157
TDS	10,580–20,000	3,500	–
Total chromium	3.0–133.0	1	0.9
Sulphide	0.0–288.0	1	1.75

Source: ETPI 1997: 19.

Pakistan are located in residential neighbourhoods, which poses a direct threat to the health of the urban population.[31]

Parikh et al. (1995) also mention several other environmental effects in their report on the Indian leather industry. These include overgrazing by cattle, the smell of rotting flesh near the tanneries, the odour of sulphide emissions from the de-hairing and ammonia emissions and flue gas emissions from the de-hairing and fleshing. The ammonia emissions resulting from de-liming cause irritation of the respiratory tract. Other negative effects of the ammonia emissions include the loss of land productivity, retardation of the germination of seeds, headaches, stomach aches, dizziness, night blindness, dermatitis and other skin disorders. Leather dust results in allergies and cancers affecting those living near the tanneries.

As in the case of textile effluents, the audits generated data enabling us to compare the effluent parameters relative to Pakistan and US EPA standards.

Table 4.5 shows that there is much more to be concerned about in the leather industry relative to the textile and clothing industry since leather production effluents far exceed both Pakistani and US EPA standards on all counts.

As in the case of all industries, the poorest are the worst affected by the pollution. First, for generations, leather-related jobs have been done by the lower castes. Second, the competition for such jobs is so intense that the manufacturers do not have to improve the dangerous working conditions. Third, the emissions affect those living around industrial sites in low-value land who have the least political power.[32]

Trade Liberalization and Export Growth in the Textile and Leather Sectors

The non-tariff barriers on trade in textiles and clothing have significantly affected Pakistan under the Multi-Fibre Arrangement (MFA) – the GATT rules that determined import quotas for various developing countries into the OECD countries, particularly the USA and those in the EU. This assertion is premised on the fact that a substantial part of its textile exports is geared towards restricted markets, and the quota utilization rates have been high. In 1994, Pakistan exported 5 per cent of yarn, 28 per cent of fabrics and 71 per cent of textile made-ups to countries that impose textile quotas under the MFA. In 1992, 86.5 per cent of Pakistan's exports to OECD countries comprised textiles and clothing. Between 1985 and 1988, the average weighted quota utilization rates for textiles exported to the US by Pakistan was 89.6 per cent; for the European market, this rate was 107.2 per cent (Ingco and Winters 1996: Tables 11 and 12).

The Agreement on Textiles and Clothing (ATC) aims to reduce non-tariff restrictions under the MFA as well as non-MFA restrictions on trade. The agreement includes the following: progressive expansion of existing quotas; integration of textiles and clothing products into GATT rules; and safeguards to deal with cases of market disruption during the transition.

The MFA-related quantitative restrictions are to be removed in three phases by 2004.[33] In each phase, importers will transfer, from the MFA to normal GATT rules, a tranche of products related to the share of these items in their total 1990 import volume. The integration into GATT rules is supposed to be implemented in three phases. In the first phase, countries were to integrate into the GATT, products from the specific list in the agreement, which in 1990 accounted for at least 16 per cent of the total volume of imports. In the second phase, beginning on 1 January 1998, products specified in the agreement that in 1990 accounted for at least 17 per cent of the total volume of 1990 imports were to be integrated into the GATT. The third phase, beginning 1 January 2002, is to integrate products in the specified list that accounted for at least 18 per cent of the total volume of 1990 imports. All remaining products are to be integrated at the end of the implementation period – 1 January 2005. A formula was developed to increase the existing growth rates for quotas of products that were under bilateral restraint. During the first phase, the growth rates were to be raised annually by not less than the growth rate established for the respective restrictions increased by 16 per cent. In phase two, the growth rates were to be the phase one rates increased by 25 per cent. In the third phase, the growth rates are to be phase two rates raised by 27 per cent.[34]

As indicated earlier, since much of the textile industry pollution is generated from the production of cloth, our focus is exclusively on cloth exports. Pakistan's future exports of cloth could be contingent on a number of factors, including the following:

1. WTO Agreement on Textiles and Clothing (ATC);
2. growth in production of raw materials such as cotton;
3. growth in manufacturing production capacity and domestic absorption; and
4. quality and exchange rate determinants of competitiveness.

Ingco and Winters (1995, Table 9) forecast the increase in Pakistani cloth exports based on the Uruguay Round agreements. Since only 28 per cent of cloth exports went to quota countries, we used the ARIMA model to forecast exports to non-quota countries. The same model has also been used to forecast exports of hides and skins, leather and footwear.[35] The results are presented in Table 4.6. While our concern is with identifying the environmental impact of export-related leather tanning (i.e. directly as leather or the leather equivalent of footwear exports), forecasts of hides and skins provide context for the overall export scenario for the leather industry discussed below.

The cloth exports forecast for Pakistan may be overstated for four reasons. First, the transitional safeguard measures against import surges have already been used by the USA, about two dozen times, against over a dozen countries.[36] Second, Pakistan faces many potential trade barriers on environmental grounds (for both textiles and leather).[37] Third, Metha (1997) pointed out that in the first phase of the ATC (1 January 1995 to 31 December 1997), developed countries did not implement the ATC clauses of 16 per cent integration of MFA into the GATT: that is, the quotas were not removed. Finally, Pakistan will face stiff competition from traditional competitors such as Bangladesh, India and China, and perhaps new ones, and so cannot take for granted access to the new market opportunities that will open up.

Tough controls on the highly polluting tanning process have contributed to a large cut in the number of tanneries in most OECD countries.[38] As a consequence, exports from LDCs such as Pakistan filled in the availability gap in these OECD countries. This probably partly explains the cumulative rapid leather export growth statistic from 1980 to 1990 of 108 per cent for Pakistan.[39] Since then, leather imports have confronted restrictions in some OECD countries based on health

Table 4.6 Benchmark and forecasts for cloth, hides and skins, leather and footwear

Product	1996–97	2004
Cloth (million sq. metres)	1,257.4	2,276.1
Hides and skins ('000 kg)	45.0	57.6
Leather (million sq. metres)	14.3	13.2
Footwear (million pairs)	8.2	8.0
= leather (millions m²)	3.01	2.94
Total leather exports (million sq. metres)	17.31	16.14

Source: Benchmark data drawn from Government of Pakistan 1998: 168–170.

criteria. For example, in 1990, Germany imposed a ban on leather treated with pentachlorophenol (a carcinogenic chemical preservative). Subsequently, several European countries imposed a ban based on the use of azo dyes.[40] Thus it is not surprising that leather export growth has tapered off for Pakistan, and the trend forecast suggests declining growth into the future.

Another reason for this is the tariff escalation used by industrialized countries. Thus, while hides and skins face zero tariffs, semi-manufacturing leather faces an average tariff of 4.8 per cent and finished goods face a tariff of 12 per cent.[41] It should not be surprising that our trend forecast shows a continued robust export growth for hides and skins. Thus it seems that, as industrialized countries have adopted cleaner technologies, they would rather import the raw materials from the South and engage in the higher value added activity themselves.[42]

Exports of leather products are slated to receive below average tariff reductions in industrial countries as a result of the Uruguay Round. The overall reduction is 18 per cent, decomposed into 11 per cent for North America and 23 per cent for Europe.[43] Thus our forecast of footwear could be biased downwards by not explicitly taking account of this tariff reduction, but not by much.

The decline in the exports of leather and footwear has occurred despite a range of export incentives provided by the government of Pakistan. These include rebates on leather product exports, duty-free imports of raw hides and skins for re-export as higher-value products and an export refund scheme for leather footwear.[44]

A more serious issue from Pakistan's perspective, however, is the immense contribution to total industrial pollution currently made by leather tanning, as suggested by Table 4.6. Anticipating and addressing the scale of the environmental threat this industry represents is critical for environmental policy.

Environmental Impacts and Mitigation Options in the Cloth Production and Leather Tanning Industries

Cloth manufacturing

SELECTION OF PARAMETERS TO COMPARE THE BASELINE INFORMATION WITH THE INCREASED POLLUTION LOAD Out of nine waste parameters for cloth manufacturing, four are within or very close to the permissible limits (pH, TDS, total chromium and phenol). The temperature is not liable to be affected with the increase in production or the effluent quantity. Sulphide is of relatively minor importance, as the generated quantity is small compared to some other parameters.

Some toxic compounds that are generated in very small quantities, such as metals, surfactants and chlorinated solvents, have also not been included in the study; we have therefore concentrated on BOD, COD and TSS.

BASELINE POLLUTION LOAD   As indicated in Table 4.7, textile effluents have high BOD, COD and TSS. Natural impurities extracted from the type of fibre being processed, along with the chemicals used for the processing, are the two main sources of pollution. Other pollution-related variables are the nature of technology and the amount of water and chemicals used in a particular manufacturing plant.

Effluents from each individual process, therefore, vary substantially. For all textile mills processing the same fibre, effluent characteristics are broadly similar but quantities may vary. For this study, the average values of the audit results have been taken as the baseline pollution level. These average figures have then been converted into pollution load per ton and per million square metres of fabric are reported in Table 4.7.

From the information in Table 4.7, it is evident that 1 ton of processed cloth produces 13.28 kg BOD, 52.08 kg COD and 70.05 kg TSS. Likewise, one million square metres of processed cloth produces 2.656 tons (2,656 kg) BOD, 10.416 tons (10,416 kg) COD and 14.01 tons (14,010 kg) TSS. The pollution load increases proportionately with the increase in production if no mitigation measures are taken.

CLEANER TECHNOLOGIES AND MITIGATION   The purpose of this exercise is to identify costs and benefits of pollution mitigation. Some

Table 4.7 Base line pollution load

Parameters	Mg/l.[1]	Kg/ton fabric	Tons/million sq. metre fabric[2]
BOD	116	13.28	2.656
COD	455	52.08	10.416
TSS	612	70.05	14.010

Notes: 1. = Effluent flow of 1,488 m3/day from 13 tons/day fabric production. 2. = Basis: 1 ton equivalent to 5,000 sq. metres, i.e. 200 GSM (grams per sq. metre).

Source: Table 4.3 and SDPI/TTSID 1995. For conversion factors from mg/l to kg/ton and tons/million sq. metres (see next sub-section).

of the cleaner technologies need to be imported, while local adaptation is possible in other cases. The SDPI's Technology Transfer Project for Sustainable Industrial Development (TTSID), has investigated various pollutants in the effluents discharged by three medium-size textile mills. Samples of effluent streams have been collected and analysed for different parameters of the NEQS (see column 1, Table 4.7). Simultaneously, flow rates of these effluent streams have also been measured and information about raw materials, process details and actual production was also collected as part of the environmental audit.

Based on this information, a baseline is available from which we can calculate the pollutants being discharged in weight per ton of production. This information is used to estimate the increased production attributable to export growth, and the proportionate rise in pollution load is calculated. We have conducted empirical exercises for two alternative scenarios: first, that of the increased pollution loads if no mitigation measures are taken, and second, that of the pollution load after installing pollution control technologies/equipment. The costs are calculated based on the pollution loads and effluent flows.[45]

Liquid waste can be reduced in both volume and concentration by a combination of internal in-plant control measures as well as external end-of-pipe treatment, discussed in Chapter 3. An SDPI/TTSID study (1995) estimated the reduction in pollution loads based on primary and secondary treatment using activated sludge technology in a 13-ton-per-day cloth-processing facility. These mitigation measures are likely to reduce the BOD level by 94 per cent, the COD level by 80 per cent and the TSS level by 98 per cent. The reduction in pollution load attained is reported in Table 4.8.

It is clear from the figures in Table 4.8 that pollution load for every

Table 4.8 Reduction in load through mitigation measures

Parameters	Present load (tons/ m. sq metres)	Total present load[1] without mitigation (tons)	Reduced load with mitigation (tons/m. sq. metres)	Total reduced load[1] after mitigation (tons)
BOD	2.656	3,339.654	0.1062	133.536
COD	10.416	13,097.078	2.0832	2,619.415
TSS	14.010	17,616.174	0.2802	352.323

Notes: 1. Based on 1,257.4 million sq. metres cloth exported (see Table 4.5).
Source: Table 4.5 and SDPI/TTSID 1995.

million sq. metres of processed cloth will be reduced as follows, if proper mitigation measures are taken:

BOD   2,656.0 – 106.2   = 2,549.8 kg (2.549 ton)
COD 10,416.0 – 2083.2 = 8,332.8 kg (8.332 ton)
TSS   1,4010.0 – 280.2 = 13,729.8 kg (13.729 ton)

In absolute terms, a reduction in pollution load can be achieved as shown in Table 4.9.

Table 4.9 Reduction in total pollution load (tons)

Parameters	Present pollution load without mitigation	Reduced pollution load after mitigation	Total reduction achieved
BOD	3,339.654	133.536	3,206.118
COD	13,097.078	2,619.415	10,477.663
TSS	17,616.174	352.323	17,263.851

With the trade-related increase in exports from 1,257.4 to 2,276.1 million sq. metres by the end of 2004, the pollution load is estimated to increase as shown in Table 4.10.

Table 4.10 Trade-related increase in pollution load

Parameter	Without mitigation (tons)	With mitigation (tons)
BOD	6,045.321	241.722
COD	23,707.857	4,741.570
TSS	31,888.161	637.762

There is therefore considerable urgency for introducing mitigation measures since without them, the export-related pollution load would increase by 81 per cent.

The cost estimates based on primary and secondary treatment using activated sludge technology for a 13-ton-per-day cloth-processing facility are reported in Table 4.11. Based on the above total mitigation cost of Rs. 38.75 million, and assuming an inflation rate of 12 per cent per year, the total mitigation cost is estimated to be Rs. 54.8 million in 1996–97 for a 13 ton/day (or 4,290 ton/21.45 million sq. metres of cloth per year) plant. This is approximately Rs. 2.55 million per million sq. metres annual capacity. Below are three exercises that provide context for policy decisions for government and industrialists.

Incremental export-related mitigation cost   Based on the above mitigation cost and the forecast increase exports of 1,018.7 million sq. metres of cloth between 1996–97 and 2004, the total estimated mitigation cost is about Rs. 2.598 billion. Thus while the incremental trade-related pollution is very high, as are the potential benefits from avoiding health and other social costs, the direct costs of mitigation are quite low in a macro perspective. Rs. 2.598 billion represents about 0.12 per cent of 1996–97 GNP.[46] Considering that textiles represent 28 per cent of total industrial sector value added, this would mean achieving sizeable benefit at very modest costs.

Total foreign exchange liabilities   The cotton chain study (Chapter 3) estimated that there are 650 units in the integrated sector, with a finishing capacity of 1,150 million square metres of finished cloth per year. As calculated above, the total mitigation cost of one million square metres of cloth is Rs. 2.55 million. Table 4.11 also suggests that the

Table 4.11 Mitigation cost estimated in 1995

Items	Local	Foreign	Total	% of total capital cost Rs. million
Civil work	6.20		6.20	16.00
Utilities and off sites	–	–	–	0.00
Plant & machinery	14.35	8.30	22.65	58.45
Inland transportation	0.20	–	0.20	0.52
Installation costs	–	–	–	0.00
Detailed engineering	–	–	–	0.00
Process design fee	–	–	–	0.00
Projects overheads	0.50	–	0.50	1.29
Contingencies	3.00	–	3.00	7.74
Sub-total	24.25	8.30	32.55	84.00
Interest during construction	4.62	1.58	6.20	16.00
Total project capital cost in million rupees	28.87	9.88	38.75	100.00

Source: SDPI/TTSID 1995.

foreign exchange liabilities are about 25.5 per cent of the total mitigation cost (1,150*2.55 = 2,932.5). Thus the total foreign exchange requirement of the country for mitigation is about Rs. 747.79 million, which is expected to increase to about Rs. 1,480.03 million if mitigation measures are adopted by 2004. The base year foreign exchange liability represents 1.5 per cent of the 1996–97 value of cloth exports.[47]

Mitigation cost as percentage of sales revenue Given government fiscal constraints, it is important to demonstrate that the mitigation costs for the industrialist are modest. The sales price of finished cloth has a large variation depending on the processing cost and its end use. For this exercise, we drew on Chapter 3 to assume an average sales price of Rs. 30. Total sales revenue from a plant of 21.45 million square metres production will thus be Rs. 643.5 million. An initial investment of Rs. 54.8 million in the effluent treatment facility for this plant forms 8.5 per cent of their sales revenue for one year.

This treatment plant would require an annual operating cost of Rs. 3.14 million. The annualization of capital costs depends on import duties, interest rate and other taxes. These will range between zero (in the case of a grant) and Rs. 7.28 million per year, as computed in Chapter 3. This yields a total annual cost ranging between Rs. 3.14 million and Rs. 10.42 million. In other words, the total annual cost of treatment facilities is between Rs. 0.15 and Rs. 0.49 per square metres of finished cloth. The total annual treatment cost thus constitutes 0.48 per cent to 1.6 per cent of sales revenues. Thus, at a micro level, the costs are once again rather modest relative to anticipated benefits reported later.

Leather According to leather industry development organizations, there are currently 526 leather tanneries in the country, most of them medium-sized. Both the benchmark and export forecast of leather and footwear are reported in Table 4.5. The pollution loads have been computed based on estimates of leather exports and the leather equivalent of footwear exports.

POLLUTION LOAD FOR CHROME-TANNED LEATHER PRODUCTION Both vegetable and chrome tanning processes are employed in the manufacture of leather. When applied to skins or hides, these produce different levels of pollution loads. In Pakistan since most of the tanneries are chrome process based, all reported pollution loads are based on this process. A comparative material balance sheet representing chrome-tanned leather from hides and skins is described in Table 4.12.

Table 4.12 Comparative material balance sheet representing chrome-tanned leather from hides and skins (kg)

	Hides 10,000	Skins 10,000
Leather	1,880	1,160
Wastewater (M^3)	120–370	1,100–2,860
Untanned solid wastes	2,700	2,274
Tanned solid wastes	2,490	1,710
Biological oxygen demand (BOD$_5$)	1,000	11,354
Chemical oxygen demand (COD)	2,500	28,386
Suspended solid	1,500	2,315
Chromium	60	66
Sulphide	100	144

Source: ETPI 1997: 30; Sadiq 1989: 45.

Biological oxygen demand (BOD$_5$), chemical oxygen demand (COD), suspended solids (SS), chromium (Cr) and sulphides (S^{2-}), are the major pollutants in tannery wastewater and hence the subject of analysis. Separate data for the annual production of leather from hides and skins are not available. However, from the data available for the year 1992–93, the leather production has been estimated to be 40.9 per cent and 59.1 per cent from hides and skins respectively.[48] The total production for 1996–97, as reported in Table 4.5, was 17.31 million m^2. Thus leather production from hides (40.9 per cent) for 1996–97 was 7.08 million m^2 and leather production from skins (59.7 per cent) for 1996–97 was 10.23 million m^2. Based on these values, the resulting pollution load for 1996–97 and 2004 have been estimated and shown in Table 4.13 and 4.14.

Data in Table 4.13 indicate that in 1996–97, a pollution load of 259.74 million kg, based on COD, resulted from the manufacture of 17.31 million m^2 of leather and the leather equivalent of footwear exported in the benchmark year of 1996–97. We reported in Table 4.4 that the total leather production for 2004 was 16.14 million m^2. Thus leather production from hides (40.9 per cent) for 2004 is expected to be 6.61 million m^2 and leather production from skins (59.7 per cent) for 2004 is expected to be 9.53 million m^2.

Data in Table 4.14 indicate that in 2004, a pollution load of 242 million kg, based on COD, would result from the manufacturing of 16.14 million m^2 of leather and leather equivalent of footwear.

Table 4.13 Baseline (1996–97) pollution loads for chromium-tanned leather manufacturing from hides and skins for leather and leather footwear exports

	Pollution load per 1,880 kg leather from hides (kg)	Pollution load per 1,160 kg leather from skins (kg)	Pollution load for 7.08 million m² leather from hides (kg)	Pollution load for 10.23 million m² leather from skins (million kg)	Baseline (1996–97) total pollution load for 17.31 million m² leather (million kg)
Untanned solid wastes	2,700	2,274	10.16	20.05	30.21
Tanned solid wastes	2,490	1,710	9.38	15.11	24.49
Wastewater (m³)	120–3,704 (av. 1,980)	1,100–2,860 (av. 245)	0.92	17.46	18.38
Biological oxygen demand (BOD_5)	1,000	11,354	3.77	100.13	103.90
Chemical oxygen demand (COD)	2,500	28,386	9.41	250.34	259.74
Suspended solids	1,500	2,315	5.65	20.42	26.07
Chromium	60	66	0.22	0.58	0.80
Sulphide	100	144	0.38	1.26	1.64

Source: Table 4.6 for benchmark for leather and footwear; Table 4.10 for pollution load, from hides and skins.

Table 4.14 Forecast (2004) pollution loads for chromium-tanned leather manufacturing from hides and skins for leather and leather footwear exports

	Pollution load per 1,880 m² leather from hides (kg)	Pollution load per 1,160 m² leather from skins (kg)	Pollution load for 6.61 million m² leather from hides (million kg)	Pollution load for 9.53 million m² leather from skins (million kg)	Forecast 2004 total pollution load for 16.14 million m² leather (million kg)
Untanned solid wastes	2,700	2,274	9.49	18.68	28.17
Tanned solid wastes	2,490	1,710	8.75	14.04	22.79
Wastewater (m³)	120–3,704 (av. 245)	1,100–2,860 (av. 1,980)	0.86	16.27	17.13
Biological oxygen demand (BOD$_5$)	1,000	11,354	3.52	93.28	96.80
Chemical oxygendemand (COD)	2,500	28,386	8.79	233.21	242.00
Suspended solids	1,500	2,315	5.27	19.02	24.29
Chromium	60	66	0.21	0.54	0.75
Sulphide	100	144	0.35	1.18	1.53

Source: Table 4.6 for forecast for leather and footwear; Table 4.12 for pollution load from hides and skins.

MITIGATION MEASURES FOR POLLUTION CONTROL Tannery effluents are regarded as a very peculiar form of polluted wastewater because they vary across tanneries both in volume as well as in pollution load.[49] As such, each tannery presents its own effluent problem. Therefore, even for a special type of leather, it is difficult to formulate a standard scheme for effluent treatment.

The methods in use for the effluent treatment may be of a physical, chemical or biological nature, used either alone or in combination. A brief account of some of these methods employed in the country has already been reported in Chapter 3. Like all other industrial wastewater treatment, the treatment's cost can be substantially reduced by adopting good in-house practices, measures related to waste reduction at source and employing more environment friendly processes/technologies as recommended by the Environmental Technology Programme for Industry for the leather sector in Pakistan.[50] The pollution removal performance of some preliminary and primary processes for tannery wastewater treatment is reported in Table 4.15.

Besides the pollution removal, other factors considered for assessing the feasibility of a process/technology to develop a treatment plant for a tannery unit include the size of the factory area, volume/flow-rate of wastewater, characteristics (qualitative and quantitative) of raw wastes and operation/maintenance requirements and cost.[51] The data described in Tables 4.16 and 4.17 are based on the findings and recommendations of ETPI (1997) from an environmental audit of three tanning units in Pakistan.

It is evident from the data in Table 4.16 that, with the ETPI-recommended treatment technology, the (1996–97) COD/BOD pollution load of the effluents after the treatment would be reduced from 363.64

Table 4.15 Pollution removal in percentages and performance of preliminary and primary processes for tannery wastewater treatment

	BOD	COD	SS	S	Cr
Screening equalization in	5	–	7	–	–
Holding basins	–	–	–	–	7
Sedimentation	45	60	70	15	53
Electrocoagulation	56	–	83	32	65
Chemical coagulation	62	–	87	–	98
Catalytic oxidation	–	–	–	90	–

Source: Sadiq 1990: 66.

Table 4.16 Possible reduction in baseline (1996–97) and forecast (2004) pollution loads of export-related tannery wastewater

Pollutants	Estimated reduction (%)	Baseline (1996–97) pollution load (million kg)		Forecast (2004) pollution load (million kg)	
		Untreated	Treated	Untreated	Treated
Water (m³)	18.50	18.38	14.98	17.13	13.96
Suspended solids (kg)	80.00	26.07	5.21	24.29	4.86
(BOD₅) / COD (kg)	65.00	363.64	127.27	338.80	117.58
Sulphide (kg)	56.50	1.64	0.71	1.53	0.67
Chromium (kg)	90.00	0.80	0.08	0.75	0.07

Source: For estimates of percentage pollution reduction see ETPI 1997; for pollution loads for 1996–97 and 2004 see Table 4.11 and Table 4.12.

million kg to 127.27 million kg. For the year 2004, the reduction would be from 338.80 to 117.58 million kg. Table 4.17 provides cost estimates for such mitigation and for chromium recovery.

The data given in Table 4.17 indicate that the operations and maintenance cost for treating 14.98 million m³ wastewater (Table 4.15) and reducing the volume of wastewater through in-plant control measures, for the manufacture of 17.31 million sq. metres (Table 4.6) of leather is Rs. 119.78 million. The total treatment cost to achieve the desired reduction in pollution, including the annualized cost (8.96 million Rs.) of the PTP (primary treatment plant), is Rs. 128.74 million.

Again, the data in Table 4.17 indicate that the operational and maintenance cost of the CRP for the recovery of 0.80 million kg waste chromium produced during the manufacture of 17.31 million sq. metres leather (Table 4.6) is estimated at Rs. 5.33 million. The total chromium recovery cost for 95 per cent recovery (0.76 million kg) of waste chromium, including the annualized cost (Rs. 0.1999 million) of CRP would be Rs. 5.53 million.

To summarize:

- Cost of primary treatment of 14.98 million m³ wastewater = Rs. 128.74 million.
- Cost of chromium recovery (0.76 million kg) from wastewater = Rs. 5.53 million.
- Total cost for wastewater treatment and chrome recovery for pollu-

Table 4.17 Cost estimates for primary treatment plant (PTP) and chemical recovery plant (CRP) for export-related tannery effluents (million rupees)

PTP for a tannery with average production load 12,000 kg hides/day (average wastewater discharge approx. 2700 m³)			CRP for a tannery with average production load of 12,000 kg hides/day (average recoverable chromium load approx. 72 kg)		
Capital cost	Annualized cost	Operations and maintenance cost/annum (17.5% of capital cost)	Capital cost	Annualized cost	Operations and maintenance cost/annum (17.5% of capital cost)
45.00	8.96	7.88	1.00	0.199	0.175

Note: The annualization was done based on a ten-year plant life and a 15 per cent interest rate, which is close to the current borrowing rate.

Source: ETPI n.d.

tion load generated from the manufacture of 17.31 million sq. metres
leather (1996–97 leather exports) = Rs. 134.27 million.
- The market value @ Rs. 45/kg of 0.76 million kg of recovered
  chromium = Rs. 34.20 million.
- Thus the net cost of mitigation (total cost minus value of recovered
  chromium) = Rs. 100.07 million.

The numbers above can be used to provide macro and micro contexts
as in the case of cloth mitigation costs.

- *Incremental exports related mitigation cost* In the case of leather, the
  incremental export-related mitigation costs are not applicable since,
  based on past trends, we project a decline in 'gross' leather exports.
  Thus, we calculated the total macro mitigation costs of putting all
  export-related leather wastewater through a primary treatment plant
  in the base period (1996–97). The cost of achieving mitigation would
  be Rs. 134.27 million or 0.0064 per cent of GNP. If the value of
  chrome recovery is netted out, the mitigation cost would have been
  0.0048 per cent of GNP. These costs of mitigation are much lower
  than those for cloth, since clean technology is locally available. In
  any case, these results strongly reinforce the finding emerging from
  cloth exports that the macro mitigation costs are very modest.
- *Total foreign exchange liabilities* Since the technology used and recommended is indigenous, there would be no capital cost-related foreign
  exchange liability resulting from the mitigation.
- *Mitigation cost as per centage of export revenue* Since the primary treatment plant is anticipated to serve several manufacturing plants at the
  same time, we have estimated the mitigation costs for the producers
  as a whole rather than for an individual unit as in the case of cloth
  production. The total export revenue from footwear and leather
  export was Rs. 11,336 million in 1996–97, and so the mitigation costs,
  net of chrome recovery, would have amounted to 0.88 per cent of
  the export revenue of the industrialists. Thus, at a micro level, the
  mitigation costs are once again rather modest relative to anticipated
  benefits.

Benefits

As explained in the conceptual framework, the benefits are calculated
indirectly by attributing a share of the aggregate costs of pollution to
cloth and leather production. Brandon (1995: 18) calculated the total
annual costs to range from a low estimate of 2.6 per cent to a high
estimate of 5.0 per cent of GDP. The costs include those resulting from

urban air pollution, water pollution (production and health impacts) and soil degradation, and also those resulting from rangeland degradation, deforestation and tourism. If we confine ourselves to the first three, the high-end cost amounts to $1,849 billion for 1992 or 4.11 per cent of GDP.[52] Since Da Silva and Qizilbash (1998: 18–19) argue that, for various reasons, even the high estimate is an understatement, we work with the high estimate. If we assume that the high-end cost for 1996 is the same as a percentage of GDP, this would amount to $2,083 billion.[53]

The challenge is to determine how much of the estimated pollution above can be attributed to cloth and leather exports. Only a very broadly illustrative 'back-of-envelope' exercise is possible. If we assume that the bulk of the pollution above can be attributed to industry and that each industrial sector contributes roughly equally to the pollution in proportion to their share in the value of production, the pollution contribution of textiles and clothing and leather and leather products would be $548.8 million and $60.0 million respectively.[54] Exports were valued, for the relevant year, as 61.4 and 64.1 per cent of the value of production for textiles and clothing and leather and leather products respectively.[55] Thus, as a rough approximation, we could attribute $ 337.0 million and $38.5 million as the pollution contribution of the exports of textiles and clothing and leather and leather products respectively.

If the mitigation measures suggested earlier are adopted and the upper-bound estimates of 91 per cent and 66 per cent of the existing pollution in cloth and leather production are reduced, the benefit of such reduction would be roughly (.91*337) $306.7 and (.66*38.5) $25.4 million respectively. This amounts to 0.5 and 0.04 per cent of GDP in 1996.[56] Thus these benefits compare very favourably to the costs that were estimated to be 0.12 and .0048 per cent of GDP.

Stakeholder Dialogues and Policy

The purpose of this section is to draw on our findings, stakeholder consultations and the literature to derive policy lessons in the Pakistani context. From discussions with officials in the relevant ministries, we found that there are no institutions in Pakistan that are currently dealing with the subject of trade, environment and sustainable development. Consultations with industry representatives revealed that, besides the need for ISO 9000 certification, Pakistani exporters are not aware of the relevant environmental policies being adopted by OECD countries.

One could rely on information flows regarding standards and environmental policies via normal market channels. However, in the case of surgical goods exports to the USA, by the time such information became available to exporters, it was already too late. The surgical goods industry had to face a ban of several years before government intervention and support enabled the required standards to be met. The same is true for shrimp industry exports to the EU. Thus the government needs to be proactive in acquiring information about environmental standards and passing this information on in a timely manner to industry, working closely with the various industry chambers. The economic case for this derives from information as a public good that confers positive externalities.

There are various government institutions though which awareness regarding environmental standards and regulations could flow to the export sector. These include the Ministry of Commerce, Ministry of Industries, Ministry of Environment, Local Government and Rural Development and the Export Promotion Bureau. The lack of such information is resulting in a loss of markets. The Ministry of Commerce may consider including a trade and environment section in the cell that deals with the WTO to draw on the relevant expertise from the other ministries.

The policy development processes relating to trade and environment in Pakistan are handicapped due to a lack of coordination and information-sharing among the relevant agencies. Trade polices are developed and implemented by the Ministry of Commerce and environmental policies by the Ministry of Environment, Local Government and Rural Development. While mechanisms exist in principle to deal with inter-agency coordination in a general sense, a specific mechanism of joint work agenda for the trade and environment section of the Ministry of Commerce will facilitate coordination. This should enable the Pakistani exporters to avoid standards-related market loss and to target green consumers.

The WWF (1997) and the OECD (1996) provide excellent policy prescriptions and examples of policies adopted to meet the challenge of environmental and health standards. A particularly relevant example for Pakistan cited by the WWF (1997: 17) and Jha (1997) relates to the response of the government of India to the challenge of meeting standards regarding dyes. The textile committee of the government prepared a comprehensive list of market regulations and acceptable alternatives to banned dyes. This information was then systematically disseminated, although SMEs were hard to reach. Eleven laboratories were also established to test for the azo dye level in products on a cost basis to ensure that standards were not being violated (see also Chapter 6).

Pakistan now has a rigorous environmental policy in place. The 1997 Environment Protection Bill, which emerged from a consultative process, was enacted in December 1997. One key feature of the Bill is that it requires manufacturing companies to conform to National Environmental Quality Standards (NEQS) or else pay a pollution charge. In July 1999, the Pakistan Environment Protection Council, the highest executive organ responsible for implementing the Environment Protection Bill, met for the first time after the Bill was enacted. In a very positive development, it set January 2000 as the date for the implementation of the NEQS. The pilot phase for self-monitoring and reporting via computer software began on schedule on 1 January 2000. Thus companies have an incentive to put environment management systems in place.[57]

Our aim in showing the likely environmental impact of exports is to persuade policy-makers of the importance of effective implementation of the NEQS (for details of implementation, see Chapter 5). The exercise, which indicates the modest costs of mitigation by using cleaner technologies at both micro and macro levels, also indicates to both business and government the feasibility of adopting cleaner technologies and the likely trade and environmental benefits of doing so.

Summary

Our research shows that there are good reasons for poor countries to want to meet the environmental standards being imposed by rich countries, because the benefits of doing so for them exceed the costs. This argument is based on several premises. First, meeting environmental standards such as the ISO 14000 can ensure efficiencies and economies within the firm. Second, these standards have built into them a process of quality controls and efficient management, and these may go a long way to winning and retaining export markets. Third, meeting environmental standards also represents a win–win scenario on a macro-economic level, since a cleaner environment would lead to a reduction in health care costs, health-related productivity losses, health-related working days lost and health-related livelihood losses.[58] Fourth, from a social justice perspective, this saving gets more weight, since the poor are the most vulnerable to environmental depredations. Fifth, our research into cloth production and leather tanning shows that, contrary to the view held in the South that the costs of mitigating environmental damage are very high, in fact mitigation costs are quite modest at both macro and micro levels.

The objective of the research described in this chapter was to estimate the increase in exports of cloth and leather and footwear, based on the Uruguay Round Agreement on Textiles and Clothing (ATC) and past trends, and identify the associated pollution and the benefits and costs of pollution mitigation. Textiles and leather are among two of the most polluting industries and, within these industries, producing cloth and tanning leather are the most polluting processes. We selected the textile and leather industries because of their economic significance and their pollution impact. The textile industry ranks as number one in terms of exports, value added and employment. Leather ranks second in terms of exports and, while it is not as significant in terms of value added or employment, it is the most polluting of all the industries.

We estimated the export-related environmental impact of cloth and leather. Following that, we assessed the mitigation impact of using cleaner technologies in terms of reducing the scale of pollution and then assessed the cost of mitigation. One way of building a strong case for mitigation is to demonstrate that these industries are highly damaging to the environment and to human, plant and animal life. Ideally, one ought to quantify the cost precisely in rupee terms. A reduction of such cost thus becomes the benefit of mitigation that can then be compared to the monetary cost of mitigation. Unfortunately, since cost quantification is difficult, we have instead documented the environmental cost and indicated how this is likely to increase due to the export-related increase in production.

The main finding of this research is that, at current emission rates, the pollution impacts of the exports of cloth and leather and footwear are very large. However, the mitigation costs at the macro level of reducing the pollution load by up to 91 per cent for cloth production and 66 per cent for leather tanning are much lower than commonly considered to be the case in the South.

For textiles, BOD, COD and TSS are the main parameters for which current emissions are above local and international standards. The chemicals used in the textile industry are very toxic and corrosive, and prolonged exposure poses a health risk. Cotton dust is a health hazard since it can cause respiratory diseases. Other problems include a pernicious odour and smog. The main problem is caused by liquid effluents that are pumped untreated into drains that enter fresh water flows. This is not only a nuisance aesthetically, but also threatens aquatic life and the use value of the water. Metals and compounds such as chromium and phenol are carcinogenic, and dyes such as azo are both carcinogenic, and allergy-inducing. These effluents also pose a threat to inland and coastal fisheries, and seepage into the water table means an entry of toxic chemicals into the soil and food chain.

For leather, the pollution load currently far exceeds national and international standards on all parameters. Leather is in this respect an even more hazardous industry. In addition to the problems of liquid effluents indicated for the textile industry, solid wastes contain chromium residues that can cause perforations and bronchial carcinoma with prolonged exposure. Poultry feed manufacturers often buy wastes, and this can result in the entry of chromium in the food chain. Tests have shown chromium residues in poultry feed. The chromium and other metals in solid wastes also adversely affect plant growth. The hydrogen sulphide formed by the presence of sulphide in the effluent is highly toxic. Ammonia emissions cause irritation of the respiratory tracts. Other problems include headaches, stomach aches, dizziness, night blindness, dermatitis and other skin disorders. Leather dust causes allergies and can be carcinogenic. Research shows serious problems of such contamination in Korangi and Charsadda. Along the Karachi coast, tanneries contribute 10–15 per cent of the total pollution. In the Punjab, prime agricultural land is being contaminated and the crop yield adversely affected.

Using an ARIMA model, we forecasted exports of leather and footwear based on past trends, drew on a World Bank forecast for the increase in cloth exports due to the Uruguay Round ATC and combined this with an ARIMA forecast of cloth exports to non-quota countries. Between 1996–97 and the end of 2004 cloth exports could be expected to rise by 45 per cent, and the corresponding increase in pollution load is calculated to be 81 per cent. Leather exports are expected to decline, so one can expect a 7 per cent lower pollution load generated by leather tanning without mitigation measures. If mitigation measures are adopted, both in plant and external, up to 91 per cent of the emissions from cloth and 66 per cent of the emissions from tanning could be reduced.

The costs of such measures in 1996–97 at a macro level would have been Rs. 2.598 billion for textile processing, which amounts to 0.12 per cent of GNP in 1996–97. The foreign exchange liability for this year would have amounted to Rs. 749.79 million, or 1.5 per cent of only cloth exports in 1996–97. More important, given government fiscal constraints, on a micro level the cost to industrialists for mitigation in a plant with a 21.45 million square metres production capacity would have been a maximum of Rs. 10.42 million, or 1.6 per cent of its sales revenue. For the leather industry, on a macro level the net mitigation cost (after subtracting the value of chromium recovery) in 1996–97 would have been 0.0048 per cent of GNP, and the mitigation cost to exporters of leather would have been 0.88 per cent of their export revenue. These mitigation costs are even lower than for cloth production, since clean production technology is locally available. In view of the negative effects of pollution generated by these industries, as indicated in the preceding paragraphs, these mitigation costs seem modest indeed. This is contrary to the view expressed in the literature that the costs of establishing and operating clean technology are very high.

Our stakeholder dialogues indicate that currently industry is inadequately informed of the rapid developments on the trade and environment interface. There is little awareness of standard-setting that is currently under way in the OECD or of how competitors are positioning themselves. Often the market provides such information, but sometimes too late, as happened in the case of Pakistani exports of surgical goods and shrimps. Since information is a public good that confers positive externalities, the Ministry of Commerce, Industries, Environment and the Export Promotion Bureau should be proactive and invest resources in the relevant information generation. The private sector would have an incentive to restrict information dissemination to recover private costs rather than encourage wide dissemination to maximize social gain. This is a classic case for state provision.

A section dealing with trade and the environment in a larger WTO cell in the Ministry of Commerce, with the relevant expertise drawn from all three ministries, may work well. Such a cell could then work closely with the various industry chambers and ensure that Pakistan does not lose markets on account of non-compliance with environmental standards and gains green niche markets. The response of the Textile Committee of the Government of India to the ban in OECD countries on azo dyes is particularly instructive.

The timing is very opportune for the government to work actively with industry and civil society to pursue an environmental and sustainable development agenda and at the same time reap the dividends of export promotion this will bring. The National Environmental Quality Standards (NEQS), which are part of the 1997 National Environmental Protection Act, are now being implemented. Industry has been involved in the process of standard-setting, has agreed to paying a pollution charge for pollution in excess of the NEQS via an enforceable process of self-monitoring (as in the case of taxation) and has even agreed to the amount of the charge. The Ministry of Commerce, Industries and the Environment can strategically provide the necessary information as this process gets under way.

As pointed out earlier, cleaner production in Pakistan may mean more exports, but it also represents an important step in the direction of sustainable development that can be viewed to be about justice for current and future generations. While the impact of poverty on the environment is often mentioned, less attention is paid to the poverty-inducing aspects of environmental degradation via the loss of access to resources for livelihood and the loss of health, productivity, working days and jobs.

Notes

1. Many thanks are due in particular to Aaron Cosbey for extensive and valuable comments, and also to Shaheen Rafi Khan. This work was initiated as part of the IISD/IUCN/IDRC Project on building capacity for trade and sustainable development in developing countries. This support, and follow-up support by RING that led to a publication in *Environment and Development Economics*, is gratefully acknowledged. This chapter draws on the above research.

2. Haider Ghani works for the Aga Khan Foundation in Pakistan, while the remaining authors work for the Sustainable Development Policy Institute.

3. Brandon's (1998: 13) upper-bound estimate for the cost of inaction in dealing with environment degradation in Pakistan was $4.36 billion in1996, over half of Pakistan's export earnings in that year.

4. Mehta 1997.

5. Cloth is the most polluting product in the textile industry, and tanning the most polluting process in the leather industry.

6. Ingco and Winters 1996: 12.

7. There are no large textile plants to speak of. For more details on textile plant selection, see SDPI/TTSID 1995, and for leather plant selection, see Khwaja et al. 1995 and EPTI 1997.

8. More details on the technology are reported below.

9. See Chapter 3 for more details on the environmental and health impacts of pesticides.

10. For a conceptualization of commodity chains, see von Moltke et al. 1998: 25–65.

11. By pesticides we mean insecticides (predominantly), nematicides, herbicides, defoliants and desiccants.

12. Not all were persuaded by *Silent Spring* and the literature it spawned. Avery (1994: 89) argues that such argumentation stems from 'an almost mystical belief that manmade chemicals are more dangerous than "natural" chemicals'. The latter, such as caffeic acid, limonene and hydrazines, are in various foods and ingested in much larger quantities than pesticide residues. Also, natural chemicals prove to be as dangerous as the synthetic variety in experiments on rats. By implication, he argues, the human body is capable of handling the 'small carcinogenic insults' resulting from pest residues.

13. This is more likely to be the case for aerial spraying. Aerial spraying was used on only 1.6 per cent of total cropped area in 1992–93. Ground plant

protection in 1991–92, the latest year for which data were available, was about 20 per cent of total cropped area. Government of Pakistan 1995: 154–8.

14. See Carr-Harris and Dudani 1992: 10, 14.

15. This involves mixing in material difficult to detect but cheaper than the actual ingredients, including water, and hence diluting the pesticide's efficacy.

16. Most dramatic is an account of 194 cases of endrin poisoning in Tala-gang, Attock (Jabbar and Mallick 1994: 15). Seventy per cent of the cases were among minors between one and nine years, and in all 19 people died. Carr-Harris and Dudani (1992: 9–11) document pesticide poisoning cases in India and report 3,029 known deaths occurring in 1990–91. Sadhu (1993: 22) cites a WHO study claiming that about half a million people in the world are poisoned each year and about five thousand of these people die.

17. See Chapter 3.

18. Government of Pakistan, 1998a: 155.

19. von Moltke et al. 1998: 134.

20. The estimate of cotton production share of fertilizer use is based on a conversation with the cotton commissioner.

21. Government of Pakistan 1998a: 138.

22. The most important chemicals for the textile industry are sulphides, phenols and chromium.

23. BOD is a measure of biodegradable materials that consume dissolved oxygen during microbial utilization of organic materials. High BOD or COD reduces dissolved oxygen in receiving waters, and thus affects aquatic life adversely, releasing obnoxious odours and toxic sulphides, and killing most aerobic organisms including fish.

24. COD is a measure of non-biodegradable materials in an effluent or water body, that react with dissolved oxygen. Its effects are similar to those of a high BOD.

25. Suspended solids increase turbidity, reduce light penetration in the water and restrict plant production. They also settle in the water and destroy spawning grounds, plug fish gills and destroy breathing organisms. High TSS can make water unfit for irrigation or drinking purposes.

26. WTO 1997: 52.

27. ETPI 1997; Khwaja et al. 1995; Nasreen and Khwaja 1997.

28. Suresh and Krishna 1983: 63; Khwaja and Nasreen 1989.

29. Beg et al. 1990: 431.

30. Srinivas et al. 1984: 314.

31. Sadiq 1989: 61.

32. Taken from the internet site Trade and Environment: South Asian Cases, 'Leather Production in Pakistan', hhtp://www.american.edu/mandala/TED/HP242.HTM.

33. This paragraph is based on Cai et al. 1997: 17.

34. GATT 1994.

35. Obviously, the ARIMA model is not capable of picking out export fluctuations such as those resulting from economic events such as the 'Asian Contagion', Thus this model implicitly assumes a return to the trend line. This is adequate for our purpose, since we are concerned only with the terminal year export value.

36. Cai et al. 1997: 14.

37. Ibid.: 31.

38. Robins and Roberts 1997: 21.

39. Government of Pakistan 1998: 169.

40. CBI/CREM 1998: 10–11.

41. WTO 1997: 51.

42. It would appear that since tariff escalation results in more of the leather tanning taking place in industrialized countries that use cleaner technologies, the global pollution level is lower and leather-exporting developing countries also benefit from lower pollution. Brazil, however, took up the issue of tariff escalation and argued that if developing countries are denied higher value added production due to such escalation, they also have fewer resources and hence less ability to adopt cleaner technologies (WTO 1999: 17).

43. Cai et al. 1997: 17.

44. hhtp://www.american.edu/mandala/TED/HP242.HTM.

45. We can also calculate a progressive decrease in pollution load if mitigation measures are taken in more than one step.

46. The GNP statistic at factor cost (average for 1995–96/1996–97) is taken from Government of Pakistan 1999: 14.

47. The cloth export statistic is taken from Government of Pakistan 1999: 88.

48. *Leather Directory 1994*, Ministry of Industries and Production, GOP, Islamabad.

49. Beg et al. 1990.

50. EPTI 1997.

51. Sadiq 1989.

52. The current GDP at factor cost for 1991/92 and 1992/93, available in Government of Pakistan 1999: 13, were averaged and divided by the average exchange rate for the same two fiscal years (p. 103).

53. As in note 52 for 1995/96 and 1996/97.

54. Data on value of production, based on the latest *Census of Manufacturing Industries 1990–91*, were available in Government of Pakistan 1999: 33. These calculations understate pollution, since textile and clothing are among the dirtiest industries. However, there is an overstatement of pollution from these sources since we are attributing all the pollution to industry.

55. Government of Pakistan 1999: 87–8.

56. See note 43.

57. At the time this research was being conducted, Pakistan was subsidizing ISO 9000 series certification. Via awareness-raising and lobbying, this subsidy has now been extended to the ISO 14000 series.

58. These benefits would be forthcoming if the pollution in question were domestic. However, if the standards were concerning greenhouse gas emissions, few of these benefits would be achieved domestically.

Bibliography

Ali, M. and A. Jabbar (1992), *Effect of Pesticide and Fertiliser on Shallow Ground-water Quality*, Islamabad: Pakistan Council of Research in Water Resources.

Avery, T. D. (1994), 'Do pesticides accumulate in the environment, posing a growing risk of cancer and other diseases?', *C. Q. Register*, Vol. 4, No. 4.

Beg, M. A., S. N. Mahmood and S. Naeem (1990), 'Pollution due to tannery effluents in the Korangi industrial area, Karachi', *Pakistan Journal of Science and Industrial Research*, Vol. 33, No. 10.

Bharucha, V. (1997), 'The impact of environmental standards and regulations set in foreign markets on India's exports', in V. Jha, G. Hewison and M. Underhill (eds), *Trade, Environment and Sustainable Development: A South Asian Perspective*, London: Macmillan.

Brandon C. (1995), 'Valuing environmental costs in Pakistan: the economy-wide impact of environmental degradation', South Asia Region Internal Discussion Paper, Washington, DC: World Bank.

Box, C. E. P. and G. M. Jenkins (1976), *Time Series Analysis: Forecasting and Control*, San Francisco: Holden Day.

Cai, W., M. Isolda, G. Guevara and C. Hamilton (1997), 'Pakistan and the Uruguay Round: issues, implementation and impact', Centre for Trade Policy and Law, Norman Paterson School of International Affairs, Carlton University, No. 44, Occasional Paper in International Trade Law and Policy.

Carr-Harris, J. and A. T. Dudani (1992), *Agriculture and People: Eco-Health Hazards of Chemical Based Agriculture and Proposed Techniques for Sustainable Farming*, New Delhi: South–South Solidarity.

CBI/CREM (Center for the Promotion of Imports from Developing Countries/Consultancy and Research for Environmental Management) (1998), *Environmental Quick Scans: Textiles: A Trade Related Orientation on Environmental and Health Issues Relevant to Exporters to the EU*, Rotterdam.

Carson, R. (1962), *Silent Spring*, Boston: Houghton Mifflin.

Chaudhry, I., M. Samee, S. Zia and T. Husain (1998), 'Sustainable cotton production, trade, and environmental impact: policy issues and options for Pakistan', Islamabad: Sustainable Development Policy Institute for the World Wildlife Fund.

Colburn, T. (1994), 'Do pesticides accumulate in the environment, posing a growing risk of cancer and other diseases?', *C. Q. Register*, Vol. 4, No. 4.

Conway, G. R. and J. N. Pretty (1991), *Unwelcome Harvest: Agriculture and Pollution*, London: Earthscan.

Da Silva, Y. and M. Qizilbash (1998), 'Environmental evaluation and accounting: the case of Pakistan's forests', Islamabad: Sustainable Development Policy Institute Working Paper Series No. 29.

ESCAP (1993), 'Proceedings of the Regional FNDINAP Seminar on Fertilisers and the Environment', Chiang Mai, Thailand, September 1992.

ETPI (1997), 'Environmental report of the leather sector', draft final report, Environmental Technology Programme for Industry, Karachi.

— (n.d.), 'Responding to the environmental challenge: Pakistan's leather industry', Karachi.

GATT (1994), 'Trade and the Environment', *News and Views from GATT, Symposium on Trade, Environment and Sustainable Development*, TE 008.

Gianessi, L. (1993), 'Why chemical free farming won't work', *Consumer Research*, Vol. 76, No. 12.

Government of Pakistan (1994), *Annual Fertiliser Review 1993–94*, Islamabad: Planning and Development Division, National Fertiliser Development Centre.

— (1995), *Agricultural Statistics of Pakistan 1993–94 and 1996–97*, Islamabad: Ministry of Food, Agriculture and Livestock, Economic Wing.

— (1996), *Foreign Trade Statistics*, Karachi: Federal Bureau of Statistics.

— (1997, 1998, 1999), *Economic Survey 1996–97, 1997–98 and 1998–99*, Islamabad: Finance Division, Economic Advisor's Wing.

— (1998a), *Agricultural Statistics of Pakistan 1996–97*.

— (1998b), *Economic Survey 1997–98*, Islamabad, Finance Division, Economic Adviser's Wing.

Harte, J., C. Holdren, R. Schneider and C. Shirley (1991), *Toxics A to Z: A Guide to Everyday Pollution Hazards*, Berkeley: University of California Press.

Ingco, M. and D. Winters (1996), 'Pakistan and the Uruguay Round: impact and opportunities: a quantitative assessment', Background Paper for Pakistan 2010 Report, Washington, DC: International Economics Department, Trade Division.

IISD and WWF (1997), 'The cotton industry: towards an environmentally sustainable commodity chain', report prepared for the workshop on Cross-National Environmental Problem Solving, School of International and Public Affairs, Columbia University.

Jabbar, A. and S. Mallick (1994), 'Pesticides and environment situation in Pakistan', SDPI Working Paper Series No. 18.

Jha, A. (1997), 'Protection of the environment, trade and India's leather exports', in V. Jha, G. Hewison and M. Underhill (eds), *Trade, Environment and Sustainable Development: A South Asian Perspective*, London: Macmillan.

Jensen, J. K. (1987), 'Sustainable agricultural crop rotation through integrated pest management', *Development, Environment and Agriculture*, Vol. 5., No. 1.

Khwaja, M. A., M. J. Rasool, M. Faiz and A. Irshad (1995) ' Survey of tannery and leather products manufacturing units in NWFP', Environmental Protection Agency, Government of NWFP.

Khwaja, M. A. and S. Nasreen (1989), 'Studies on effluents from tanneries/ leather industries around Charsadda–Peshawar Areas (NWFP)', *Proceedings, Fifth National Chemistry Conference.*

Lalonde, B. and L. Chabason (1994), 'Towards an ecological injection into international trade', Paris: Eco-Development.

Mehta, P. (1997), 'Textiles and clothing – who gains, who loses, and why?', CUTS/CITEE Briefing Paper No. 5, Jaipur, India.

von Moltke, K. et al. (1998), 'Global product chains: Northern consumers, Southern producers, and sustainability', *Environment and Trade*, No. 15, Geneva: UNEP.

Nasreen, S. and M. A. Khwaja (1997), 'Studies on sub-surface water quality around tanneries in some areas of North West Frontier Province (NWFP)', draft.

OECD (1996), 'The Environmental Effects of Trade', Paris: OECD.

Parikh, J. K., V. K. Sharma, U. Gosh and M. K. Panda (1995), *Trade and Environment Linkages: A Case Study of India, Indira Gandhi Institute of Development Research*, report prepared for UNCTAD.

Qutub, S. A. (1994), 'The divergence between private and environmental costs and benefits: a case study of chemical and organic fertiliser in Pakistan', presented at the Tenth Annual General Meeting of the Pakistan Society of Development Economists, Islamabad.

Robins, N. and S. Roberts (1997), *Unlocking Trade Opportunities*, New York: International Institute of Environment and Development and UN Department of Policy Co-ordination and Sustainable Development.

Sadhu, G. R. (1993), 'Sustainable Agriculture', a Pakistan National Conservative Strategy Sector Paper, Karachi: IUCN and Environment and Urban Affairs Division.

Sadiq, H. (1989), 'Tannery effluents treatment and strategy in Pakistan', *Pakistan Leather Trade Journal*, Vol. 16, No. 2.

— (1990), 'Tannery effluents treatment and strategy in Pakistan', *Pakistan Leather Trade Journal*, Vol. 17, No. 4.

Saleem, M. T., J. G. Davide, H. Nabhan and A. Hamid (1989), 'Soil fertility and fertiliser use in Pakistan with special reference to potash', *Potassium and Fertiliser Use Efficiency*, Islamabad: National Fertiliser Development Centre.

SDPI/TTSID (Sustainable Development Policy Institute/Technology Transfer for Sustainable Industrial Development) (1995), *Environmental Examination of the Textile Industry in Pakistan*, Final Report, Z5CR1TTI, Islamabad.

Srinivas, M. Farooque, G. Teekaraman and N. F. Ahmed (1984), ' Groundwater pollution due to tannery effluents in North Arcot District, Tamil Nadu', *Indian Journal of Environmental Health*, Vol. 26, No. 4.

Suresh, C. and G. Krishna (1983), *Pollution Research*, Vol. 2, No. 2.

UNCTAD (1995), *Trade, Environment and Development: Lessons from Empirical*

Studies, The Case of India, Synthesis Report, Trade and Development Board, Ad Hoc Working Group on Trade, Environment and Development, Third Session, TB/B/WG.6/Misc. 7/GE 95-53647, Geneva.

Weir, D. and M. Schapiro (1981), *Circle of Poison*, San Francisco: Institute of Food and Development Policy.

World Trade Organization (1997), *Environmental Benefits of Removing Trade Restrictions and Distortions*, Committee on Trade and Environment, WT/CTE/W/678, Geneva.

— (1999), *Trade and Environment Bulletin*, Press/TE 028, Geneva.

WWF (1997), *The Textile Trade: A Study for the Expert Panel on Trade and Sustainable Development* (by Shirpas Das), Gland.

CHAPTER 5

Adopting Environmental Standards: Pakistan's Response to Industrial Pollution

Haroon Ayub Khan and Abdul Matin Khan[1]

§ A pollution charge regime has been introduced in Pakistan to achieve industrial compliance with the National Environmental Quality Standards (NEQS). These standards, if successfully implemented and documented, would go a long way to meeting the standard requirements likely to be imposed by importing countries. The modalities for the implementation of the pollution charges have gone through a unique consultative process between representatives of industry, government, environmental NGOs and academic researchers. The consensus of all stakeholders has been to adopt a market-based approach, that is, a pollution charge or tax combined with fiscal incentives to industries, rather than to use coercive criminal procedures for ensuring compliance with NEQS. Appreciable progress has been made towards operationalizing the process and January 2000 was fixed as the date for commencing implementation. This was subsequently extended to 1 July 2001.[2] This chapter documents Pakistan's experience in formulating these critical environmental policy developments.

Background

Although they were originally promulgated by the government in 1983, there had never been a concerted effort to implement the NEQS until the Pakistan Environmental Protection Council (PEPC) was reactivated in 1993 by Asif Ali Zardari, the minister for environment and husband of the then prime minister, Benazir Bhutto. Patronage at this level provided the necessary political support for environmental concerns in the country and it was at this time that the Sustainable Development Policy Institute (SDPI) suggested the use of a pollution charge, based on the German experience in pollution control, and initiated discussions on modalities for implementation.[3]

Somewhat sceptical at first about the use of such a market-based instrument, government and private sector representatives soon came to the realization that this was perhaps the most effective and equitable way of ensuring compliance with NEQS. There are obvious obstacles in the transition to more sustainable industrial production, not least of which is the cost of new technology, lack of technical know-how or expertise, insufficient credit availability, and the already weak financial health of the industrial sector. Faced with these problems, it has been a challenging task to convince industry, especially the non-exporting sectors, to comply with new environmental standards.

Nevertheless, there has been a growing awareness in industry of the needs and benefits of going green, with special efforts on the part of organizations such as the APTPMA (All Pakistan Textile Processing Mills Association), the FPCCI (Federation of Pakistan Chambers of Commerce and Industry), OCAC (Oil Companies Advisory Committee) and the OICCI (Overseas Investors Chamber of Commerce and Industry). The increasing pressure, especially on exporting sectors, for new international production and management standards has made the private sector a little more conscious of the need to comply with environmental standards to remain competitive in international markets. While a few industries in Pakistan have undertaken some voluntary efforts to curb pollution, the passage of the Environmental Protection Act 1997 has now made the payment of pollution charges a statutory requirement for all non-conforming industrial units.

Unfortunately, however, whether there is an adequate legislative cover or not, the command and control measures usually employed by the government have often failed. The SDPI therefore advocated the need for dialogue from the very beginning. A systematic approach was needed that took into account the realities and limitations of both government and industry, and, in doing so, also built trust and confidence between them. This is necessary for government because it does not have the capacity to regulate all industry, and conversely, it is necessary for industry in order to ensure that any control measures initiated by the government are realistic and fair.

Early Interaction

The need for a dialogue was echoed in a meeting convened at the prime minister's residence in March 1996, where industry representatives appreciated the ongoing consultative process that had been adopted by the government to draft environmental legislation. It was pointed out that a similar process had not been followed for setting the NEQS and that industry was of the view that these were too stringent and impractical. As a result, a request was made that the implementation of NEQS be deferred for some years to enable industry to prepare and take corrective action. After much discussion, the government said that while deferral was not acceptable, a system could be worked out that imposed a progressive financial penalty starting at a moderate level in lieu of immediate implementation of the harsher penal clauses of the draft legislation. Such a system would be worked out, it was agreed, through a consultative process.

The Negotiation Process

In order to rationalize the NEQS and to work out detailed modalities for their implementation, the PEPC constituted on 12 March 1996 the Environmental Standards Committee (ESC) under the chairmanship of Dr Shamsh Kassim-Lakha, president of the Aga Khan University, and designated the SDPI as its secretariat. The members of the committee included representatives from the Ministry of Environment, Federal and Provincial Environmental Protection Agencies (EPAs), public and private sector corporations, industrial chambers and associations, environmental NGOs, research organizations and legal experts. The mandate of the ESC was very specific: 'to review the NEQS and suggest changes where necessary, and to recommend modalities for enforcing them'. But in order to accomplish this mandate, a multi-dimensional strategy would be necessary – one that combined a creative market-based formula with technical support to the industry and government, a mass awareness programme and an effective monitoring system. The only way to ensure success was to do all this openly and transparently.

The series of roundtable discussions that followed the establishment of the ESC is a classic example of a transparent and participatory policy-making process at the national level. In fact, it can be argued that the survival of the consultative process, despite a turbulent political climate in the country since 1996, has been the result of its participatory nature and the sense of ownership of the process acquired by all parties concerned.

Review of National Environmental Quality Standards[4]

Negotiations began with a review of the NEQS themselves.[5] Considerable objections were raised by the private sector that these had not been developed through public consultation and that, therefore, some of the standards were unrealistic. A technical committee was formed to address specific objections against each of the parameters in question. This review process resulted in the rationalization of the NEQS with input from industry. Because of the high toxicity of certain pollutants, however, industry-specific NEQS are likely to be more stringent than those currently applicable. In comparison with other developing countries with a similar industrial base, these standards are neither too stringent nor too relaxed. The intention was to begin with realistic limits in view of the prevailing conditions in the industrial sector and to have the possibility to tighten these further later on if necessary.

Concept of Pollution Charge

The concept of the pollution charge is a key element of the implementation programme. As mentioned before, it was introduced after extensive discussions with businesses, government and the private industrial sector, in response to concerns that the NEQS should not be enforced on existing industrial units through coercive criminal procedures. According to the present proposal, the charge would be calculated on the basis of a pollution load measured in pollution units.[6] The principle is that the charge should be high enough to induce industry to clean up its act. In other words, the net cost of the clean-up should be less than the pollution charge. Such a charge would ensure that those who introduce clean-up activities do not suffer relative to those who persist with dirty production methods.

It took more than two years of intense discussions and negotiations to arrive at the present formula for calculating the pollution charge. Numerous questions had to be resolved: NEQS are in terms of concentration – should the charge be levied on the level of concentration or on the quantum of pollutants in the emission? Should the charge be the same on all industries, or should a different charge be applied by type of industry? Should there be a charge on every component of NEQS, or should it be levied on key components? Should future increases of the pollution charge be subject to negotiations, or should they be imposed and agreed upon up front? Should fiscal incentives be made available to industries for clean-up activities? How would the pollution charge be collected and for what purpose would it be used? And, of course, the ultimate question: what will be the per unit rupee amount or base rate of the pollution charge?

Getting endorsement for the plan by industry as a whole was as important as reaching a consensus on the above technical questions. Industry representatives at the negotiating table were under intense pressure from their constituents to ease or postpone the financial impact of this programme, given the already adverse economic conditions being faced by the industrial sector. Sparks flew, tempers flared, but the representatives managed to convince many of the need for the pollution charge regime to address the long-term environmental damage likely to be caused by unrestrained industrial growth in the country.

Formula for Pollution Charge Calculations

A number of proposals were reviewed extensively for the determination of pollution charges. The following decisions were made: (1) the pollution charge should be equitable and simple; (2) it should ensure real progress towards making the industry environmentally friendly without jeopardizing economic growth; (3) industry should be allowed sufficient time to prepare for compliance. In addition, the ESC had agreed to the following:

- The level of pollution charge should be established through a process of negotiations.
- The level of pollution charge should initially be such that industry should feel the impact, but it should not be excessive such that the financial health of the concern is jeopardized.
- The system should be applied uniformly across all industrial sectors. Industry-specific application was not recommended.[7]

The initial proposal submitted by the SDPI recommended linking the pollution charge to the cost of effluent clean-up. Another variation of this proposal suggested linking it to the cost of environmental damage caused. In either case, a gradual increase was recommended so that industry would be induced to adopt cleaner production methods and technology over time.

An alternative proposal circulated by the FPCCI recommended levying pollution charges on the basis of pollution loads depending on the size and type of industry. Industries were categorized into three sets according to pollution treatment technologies: (1) parameters to be covered under primary and secondary treatment, (2) secondary and chemical recovery treatment, and (3) recovery and re-use technologies. The proposal divided the NEQS list according to these three categories and estimated clean-up costs for these respectively.

The proposal on which consensus was finally reached was developed jointly by the SDPI, the Federal EPA and Halger Bailly Inc., Pakistan. Based on the experience in Germany with the use of pollution charges, this programme will levy pollution charges on the basis of pollution units in excess of NEQS as determined by an agreed procedure. It was agreed that the application of NEQS and the levy of pollution charges will be applied uniformly to industry in the private and public sectors, and will eventually include municipal services as well.

Agreement on the Pollution Charge Amount

Irrespective of the formula or determination procedure applied, the base rate, or the actual rupee amount per unit of pollution, would obviously be the determining factor to ensure a transition towards cleaner production. Clearly, this was to be an immensely critical and expectedly controversial decision for all concerned. Numerous discussions took place and the ultimate responsibility of democratically arriving at the figure was placed in the hands of industry under the leadership of the FPCCI.

It took weeks of negotiations among industry representatives to establish both a pollution charge and a progressive escalation schedule.[8] This was an unprecedented achievement in Pakistan and perhaps also elsewhere in the world, whereby industry voluntarily agreed to a charge to be applied to themselves for generating pollution in excess of permissible national limits. The FPCCI task force also recommended that the pollution charge be renamed the Environmental Improvement Charge to evoke a more favourable response from industry.[9]

Monitoring

A major issue for the Environmental Standards Committee was the absence of an adequate monitoring capacity in the EPAs, and in the government more generally. Industry representatives were sceptical about the transparency and fairness of any system that relied primarily on monitoring by a limited number of overburdened and under-trained government inspectors. The government representatives also felt that the current capacity of the monitoring agencies was considerably short of the demands likely to be placed upon it. As such, there was a consensus on developing a sophisticated monitoring system that did not rely exclusively on government inspections. Such a system would begin initially by self-monitoring and reporting by the units concerned. These reports would be taken at face value, except in case of doubt. Also EPA authentication would be required for a randomly selected sample. Finally, reporting of compliance with NEQS from all industrial units would be placed in the public domain to enable independent researchers and environmental NGOs to monitor them and assess the performance of the entire system. Any entities that wilfully conceal or falsely declare the level of pollutants in their report will be open to prosecution under the harsher penal clauses of the Environmental Protection Act.

A simplified monitoring programme has been agreed on. Based on the degree of hazardousness and toxicity of emissions, industry has been divided into Categories A, B, and C in order of the pollution generated. For Category A, a monthly monitoring and reporting (M&R) frequency has been recommended for both liquid and gaseous emissions. For Categories B and C, quarterly and biannual M&R have been recommended respectively. For most of the industries, M&R of four to six priority parameters have been proposed under normal plant operating conditions. These M&R guidelines would be applicable to both the private and the public sectors and would be reviewed from time to time.

To ensure consistency in the sampling and monitoring process, the Federal EPA is undertaking measures to standardize sampling and testing procedures as well as certifying laboratories across the country that would be used for analysis. Furthermore, in order to ensure transparency, government and industry agreed to allow reputable NGOs to be present at any stage of the monitoring process.

Mode of Collection and Use of Funds

While the modalities of collection and disbursement of funds are still being worked out, the basic principles have also been agreed to after exhaustive discussions between industry representatives and the government. Ever since it was agreed that the money collected as pollution charges would be made available for environmental services to benefit industry (see Box 5.1), the private sector has been adamant that these funds must not be deposited in the national treasury, from where they are likely to be utilized for other purposes. Instead, they have strongly advocated the creation of Provincial Environmental Trust Funds (PETFs), which would be governed by a tripartite board of private sector, government and NGO representatives. Furthermore, the private sector has recommended that these funds be collected by industry associations. Such an arrangement is necessary, according to industry, to facilitate timely payments, both by the industrial units and subsequently by the Trust Funds, for any environmental services requested.

Although these arrangements received endorsement by the ESC and were formally submitted as recommendations to the PEPC, certain legal restrictions have prevented the establishment of such institutional arrangements. Article 11(2) of the 1997 Environmental Protection Act states that 'The Federal government [will] levy a pollution charge on any person who contravenes or fails to comply with the provisions of sub-section (1), to be calculated at such a rate, and collected in accordance with such procedures as may be prescribed.' The Act, however, does not specify where these funds are to be deposited or for what purpose they are to be used. It is implicit, nevertheless, that since the federal government is responsible for the collection of the pollution charge, they must be deposited as revenue of the federal treasury.

Article 9 of the Act calls for the establishment of Provincial Sustainable Development Funds (PSDFs). These funds can be utilized for 'providing financial assistance to the projects designed for the protection, conservation, rehabilitation and improvement of the environment, the prevention and control of pollution, the sustainable development of resources and for research in any specified aspect of environment; and any other purpose which in the opinion of the Board will help achieve environmental objectives and the purposes of this Ordinance' [Article 9(3)(a,b)]. The government is of the opinion that the PSDFs can be used as the PETFs suggested by the private sector. However, the anomaly is that pollution charges have not been included as one of their sources.[10] Industry representatives have consistently argued against this point, saying the PSDFs are not industry-specific and will result in innumerable complications.

Institutional arrangements, therefore, for the collection and adminis-

---

Box 5.1 Use of Pollution Charge

Money collected will be used primarily for activities that will help the abatement of environmental pollution through the following activities:

- provision of soft loans for the purchase of pollution treatment equipment,
- installation of combined effluent treatment plants in industrial estates,
- research and analysis in support of pollution abatement,
- roundtables, conferences and workshops for pollution abatement,
- provision of incentives to develop indigenous technology for pollution control, and
- training and advisory services for industry.

Source: ESC recommendations to PEPC, 20 May 1996.

Box 5.2 Status of Incentive Measures

Agreed recommendations

(a) National Development Finance Corporation may be designated as the DFI for channelling soft-term credit to industries for environmental purposes.

(b) Purchase of equipment for pollution abatement may be given the most favoured treatment, i.e. 10 per cent, with regard to import duty, sales tax, and no regulatory duty.

(c) Most favoured tax treatment may be extended to those developing indigenous technology for pollution control.

(d) The amount collected from pollution charges and other sources for the Provincial Environmental Trust Funds may be matched by proportional grants from the government.

(e) The use of the Provincial Environmental Trust Funds may be decided by the respective governing boards in accordance with the guidelines laid down in the recommendations of the Environmental Standards Committee.

(f) Provision of accelerated depreciation of anti-pollution equipment within three years for income tax purposes.

Current status (early 2002)

At the ninth meeting of the Pakistan Environmental Council (PEPC) held on 3 February 2001, industry representatives pointed out that 19 industries had made a total investment of Rs. 2,784 billion for pollution control and repeated their demand for the reduction of import duty on the procurement of anti-pollution equipment from 10 per cent to zero. They also requested an early finalization of other incentives for encouraging early compliance with NEQS. The PEPC asked the Federation of Pakistan Chambers of Commerce and Industry (FPCCI) to submit formally its proposal regarding incentives to industry to the Ministry of Environment to promote compliance with NEQS. This message was repeated by the minister of environment during the first meeting of the reconstituted NEQS implementation committee in December 2001, when industry raised the issue of financial incentives.

tration of pollution charges are being worked out. Legal advice is being sought to try to resolve this issue and preserve the use of pollution charges in the manner recommended by the ESC. Provisions are also needed to ensure an equitable participation of government, private sector and NGOs on the boards of the PSDFs to oversee and ensure the agreed utilization of the pollution charges.

Financial Incentives for Industry

Following extensive negotiations with the government in the ESC, the Pakistan Environmental Protection Council approved a detailed proposal for provision of fiscal incentives to industry for pollution abatement or compliance with NEQS. The current status of these incentive measures as reported by the Federal EPA is shown in Box 5.2.

Another obstacle identified by industry is the lack of credit availability for environmental technology or investment. Private financial institutions in the country are reluctant to provide loans for environmental projects because they do not see them as profitable. In any case, since savings from environmental investments are likely to be indirect and realized over the long term, industry is not willing to take on loans at commercial rates. With this realization, SDPI began investigating the possible establishment of green credit facilities on soft terms. Some international donors have expressed interest in such credit windows, and there may be potential to mobilize others. If suitable, efficient and effective channels are established, a case can be made for the international community to live up to their global commitments to protect, conserve and support environmental activities in developing countries. This has been the commitment made by Northern countries in numerous international conventions and treaties on the environment.

Increasing Technical Capacity

Effective implementation of this programme requires increased technical capacity in the private sector as well as in government monitoring agencies. Information about the experiences of industry in other countries will be of use as options and improvements are identified in Pakistan. EPAs need greater technical capacity and trained manpower to monitor compliance of industrial emissions with NEQS. The EPAs are currently in the process of standardizing analytical sampling and testing procedures, and draft regulations for certification of environmental laboratories have been prepared. All these efforts must be accompanied by extensive training and awareness-raising in industry, preferably by a multi-party initiative involving EPAs, relevant government departments, chambers of commerce and industry, environmental NGOs and other national or international agencies. A detailed action plan for implementing an environmental monitoring programme, including awareness-raising and training for industry, has recently been finalized by the Federal EPA and SDPI.

Two independent initiatives have extended such services to industry and government, and to facilitate compliance with NEQS. These are the SDPI's project on Technology Transfer for Sustainable Industrial Development (TTSID), and the FPCCI's Environmental Technology Programme for Industry. There have been five other smaller private sector initiatives.

The SDPI's programme on Technology Transfer for Sustainable Industrial Development, funded by the Swiss Federal Office for Foreign Economic Affairs, provided support to industry and government for the promotion of policies and practices for sustainable industrial production through five distinct components.

1. Business–government roundtables to facilitate regular consultations between the private sector and government on environmental issues.[11]
2. Supported by technical research, recommendations emerging from these consultations provided advice to the government for the development and implementation of national environmental policy.
3. The training component of the TTSID developed training materials by conducting environmental studies in selected industrial sub-sectors followed by hands-on training and workshops. This component also provided support to industry in building capacity in self-monitoring, implementation of in-plant pollution control measures, and in identification of end-of-pipe treatment options.
4. Through the information and advisory services component, the project produced information packages on environmental issues for selected industrial sub-sectors, developing directories of equipment, service and technology suppliers, and a database for the exchange of information.
5. Finally, the project also developed proposals for innovative financial mechanisms for the establishment of green credit facilities for environmental projects in industry.

A similar initiative is the FPCCI's Environmental Technology Programme for Industry (ETPI), funded by the Netherlands government. Its primary objective is to promote the use of environmentally sound technologies for the production of environmentally safe products in Pakistan's manufacturing and industrial sectors. This will be achieved by on-site training and demonstration projects for adopting measures for pollution abatement, waste management and recycling, chemical recovery, more efficient utilization of natural and/or economic resources, production and installation of instrumentation and control systems for utilizing more efficient and environmentally safe production technologies. The project has five components: the development of a user-friendly database of relevant information; institutional networking within and between key industrial institutions of the country; dissemination and communication to promote cleaner industrial production; institutional support and training to create environmental capacity within industrial chambers and associations; and demonstration projects in selected industrial sub-sectors to demonstrate the economic feasibility and environmental efficacy of environmental technologies.

The ETPI and TTSID programmes were complementary in nature and, though modest in scale, have met some of the immediate training and advisory requirements in this context. However, much more in this regard is needed.

The Role of Non-Governmental Organizations

The role of NGOs in this entire process has been a crucial one from the start and one that has been acknowledged by both government and industry. First, the PEPC's appointment of the leader of an NGO (the Aga Khan University) as president of the Environmental Standards Committee with the SDPI (another leading environmental NGO) as its secretariat supports this view. Second, the presence of NGOs has provided an openness and transparency to the negotiation process, and has allowed a balanced expression of opinions that catered to the interests of all concerned parties. Third, certain NGOs are playing an important role in raising awareness, not only of industry, but also of the public, about the importance and need for environmental conservation. Fourth, a few NGOs working in this sector, such as the IUCN and the SDPI, are making efforts to strengthen capacity of both the private sector and the government (see section on increasing technical capacity above). Fifth, NGOs are expected to have an important monitoring function in the future implementation of the programme. Sixth, the sharing of technical expertise between the private sector, government and NGOs has resulted in an unprecedented constructive partnership between these diverse entities. Seventh, several NGO leaders have been appointed to be members of the PEPC, which is the highest-level environment policy-making body in the country.

Although an appreciable role has been played by NGOs in the process so far, there are very few NGOs with the requisite technical knowledge or programmes in related areas. Just as with the government and private sector, environmental NGOs also need capacity-building. There are a large number of advocacy groups in the country that have also not been sufficiently mobilized to campaign for pollution controls on industry. The potential therefore exists for a much more involved interaction of NGOs in this area.

Achievements

The steps towards sustainable industrial production have been significant. The establishment of a transparent, broad-based, national consultative process has been instrumental in moving the programme for the implementation of NEQS as far as it has come. In fact, this experience is now being replicated at the provincial level in the implementation of the industrial development component of the Sarhad Provincial Conservation Strategy. The endorsement of the basic principles of the programme, and its simultaneous inclusion in the Environmental Protection Act, concretizes the initial move towards sustainable industrial development in Pakistan.

The Human Element

The human element was probably the most critical element that enabled the process to develop as it did. In particular, Tariq Banuri's (as executive director of the SDPI) initial leadership and sustained efforts, Babar Ali's allowing the first audit (at Packages), the support of Mahmood Ahmed and Zaffar Ahmed Khan, prominent leaders of the business community, and of Aban Marker Kabreji, country representative of the IUCN, particularly for ensuring that the IUCN law team included the NEQS document in the draft of the Environmental Protection Act, were all vital. Others who played a supportive role as concept champions include G. R. Arshad and Gulzar Firoz.

Asif Zardari's convening of the task force for the final agreement was essential, as was the support of Mohammed Imran Faruqi. Similarly, Asif Shuja (as director-general of the PEPA) played a supportive role during council meetings, without which the idea would have withered on the vine. Moeen Afzal (then secretary of finance), and Qazi Qalimullah (secretary-general, finance, later deputy chairman of planning) were also helpful. The point is that considerable intellectual and political leadership went into making the process a success.

Uphill Effort and Update

Notwithstanding the achievements, it has not been smooth sailing the whole time. Certain sectors of the strong industrial lobby are still trying to postpone implementation; government enthusiasm has been lukewarm at best; the PEPC remained inactive for a long period; economic crises and political unrest, with at least three changes in the government (including frequent changes at senior levels in the Ministry of Environment – the main government counterpart) since 1996, have made outcomes and direction of the process very uncertain; there continues to be disagreements on the means of collection and use of the pollution charge; and the Environmental Protection Act was enacted after a long delay and a hard struggle. The danger of the entire initiative being shelved at a moment's notice is still present.

One of the outstanding areas of disagreement in the negotiating process is the lack of a suitable institutional arrangement for the collection and disbursement of the pollution charge. While the government claims it is legally bound to use PSDFs for this purpose, industry insists on placing the funds in the private sector (see above). This remains a contentious issue because of the prevailing mistrust or apprehension of the private sector regarding the government's bureaucratic procedures and the urgent need for funds due to the ongoing fiscal crisis.

While the dialogue continued, and as the reality of the pollution charge regime loomed closer, sections of the private sector that were hitherto inactive in the negotiating process began to raise various objections. The most common of these was that they were not adequately consulted in the process. Other complaints included the fear that the EPAs would simply become another agency for rent-seeking and extraction and that they lacked adequate capacity anyway. Numerous concerns about the NEQS parameters and other elements of the programme were brought up. This was, in part, due to the failure of the industry representation process, and partly because of the insufficient creation of public awareness at the outset of the programme. The last ESC meeting of 6 August 1998, however, ruled out the possibility of reopening previously settled issues. Nevertheless, given the evolving nature of the process, mechanisms for dialogue or continual adjustment, when necessary, must be permanently institutionalized.

The first step in implementing Environmental Standards Committee (ESC) decisions was to convene the Pakistan Environmental Protection Council (PEPC). The Council, which is required by law to hold at least two meetings in a year, assembled after two and a half years in August 1999. This was exactly a year after the ESC has finalized its recommendations. This meeting was convened only after NGOs and other environmentalists exerted considerable pressure. The prime minister could not spare the time to chair this meeting and nominated the minister for the environment, who was the vice-chairman, to chair it.

During this meeting, the chairman of the ESC presented his final report and the details of various recommendations, which were adopted by the Council as a whole. The important decisions of the Council included approval of the revised National Environmental Quality Standards (NEQS) and implementation of a self-monitoring programme and pollution charge system from January 2000. The Council also decided that the Environmental Standards Committee (ESC) and the Experts Advisory Committee (EAC) would continue functioning to monitor the implementation process and develop industry-specific standards respectively.

Although progress has not been as fast as could be expected after a complete consensus on important issues, matters are moving in the right direction. Some of the important developments from August 1999 to date (March 2001) are briefly discussed below.

Self-monitoring and reporting programme It was agreed in principle in the ESC that industry would start monitoring and reporting its pollution levels in the manner agreed during these negotiations. A date of November 1998 was also decided. The Pakistan Environmental Protection Agency (PEPA) had to do some homework before implementing this programme, which included finalizing analytical methods and procedures and designating testing laboratories for this purpose. In the meantime, the government also felt that collecting and compiling the information in hard form would be difficult. The PEPA, with assistance from the Sustainable Development Policy Institute (SDPI), has now developed a software called the Self-Monitoring And Reporting Tool (SMART). This software will be used for reporting and compiling environmental data in soft form. The database has three different packages to be used by industry, provincial EPAs and PEPA respectively.

SMART The database is user-friendly by design and the user will be able to enter the data with the help of accompanying 'user instructions'. It has a registration module, which will automatically assign a distinctive user number to each reporting industrial unit in the country. These units will be able to transfer this data to EPAs electronically. All this information will be compiled through this database at the provincial as well as the national level. There is an in-built provision of data confidentiality to safeguard genuine business interests.

Pilot phase of the self-monitoring programme All these developments were shared with representatives from industry chambers and associations in a kick-off meeting in December 1999, and it was agreed to start a pilot phase from January 2000. During this phase, 50 industrial units from different provinces and from different industry groups took part in this programme on a voluntary basis. The SDPI worked as a partner with the PEPA in providing technical support to provincial EPAs during this phase. The FPCCI and the PEPA agreed on a list of participants and the software was dispatched to participating units. The main objectives of this phase included testing SMART, and a small-scale testing of the self-monitoring programme as a whole. Frequent interaction during this phase also strengthened the government–business relationship.

Information package for industry In addition to the 'user instructions for SMART', the government also put together a set of important documents that were sent to participating units as a consolidated information package. This package contained guidelines on the self-monitoring and reporting programme, sampling methods and analytical procedures, the Pakistan Environmental Protection Act 1997, draft Revised National Environmental Quality Standards, a list of recommended laboratories and addresses of resource organizations.

Revised NEQS The Expert Advisory Committee, in consultation with industry representatives and some NGOs, reviewed different NEQS parameters and identified twelve parameters for revision: ten for liquid effluents and two for gas emissions. The ESC and the PEPC approved the revised NEQS and the government notified them.

Rules and regulations During this period, the government also prepared seven sets of values and two sets of regulations pertaining to sustainable industrial development draft rules and regulations after stakeholders' consultation. The government is now in the process of notifying these rules and regulations to support the implementation process.

Update At the ninth meeting of the PEPC held on 3 February 2001, the council was informed that 371 industrial units belonging to 19 industrial sectors had invested Rs. 2.7 trillion towards pollution control. The PEPC confirmed that the levy of pollution charge would begin on 1 July 2001. Industry committed to have an investment of about Rs. 800 billion annually to contribute towards pollution control and this sum will be raised to Rs. 1.3–1.7 trillion annually (industry reiterated its request for duty free imports of 'anti-pollution' equipment). Further, industry will provide the federal and provincial EPAs with time-bound achievable targets via their associations. Also, their associations will establish environmental committees to oversee the performance of their member units with regard to conformity with the NEQS.

Lessons Learned, Recommendations and Challenges

All initiatives of this nature need to start with sound intellectual leadership and then a forceful mobilization of concept champions at the highest political level. In Pakistan, such intellectual and mobilizing leadership was very effectively provided by Tariq Banuri and several concept champions were successfully mobilized.

Following this, perhaps the most important lesson learned has been the usefulness and effectiveness of legitimizing the participatory policy-making process. The survival of the initiative, despite all sorts of potentially disruptive internal and external factors, has followed from this. The participation of a wide cross-section of the stakeholders permitted a broader understanding and greater sense of ownership in the design of the programme. Just as in any participatory consensus-building exercise, it has taken time and a great deal of negotiation at the highest levels. It was also recognized that such an initiative would not have been possible with a narrow view of operational modalities, but that a much more integrated programme was needed that included institutional support services such as information and advisory services, technical advice on the formulation and monitoring of standards, establishment of innovative financial instruments, capacity-building and regulatory and legislative support.

A neutral, business–government roundtable forum (the Environmental Standards Committee) was also necessary to provide balanced representation and unbiased mediation and to ensure the full transparency of the process (in this particular instance the ESC chairperson, Shamsh Kassim-Lakha, provided the essential and outstanding facilitating role). This forum had to be at a sufficiently high level to include key decision-makers but structured in such a way as to minimize debate on technical details that would distract their attention (and time) from making important policy choices. The ESC would, therefore, regularly form specific technical sub-committees to investigate options to facilitate the policy-making process. Once again, since representation by all stakeholders was allowed on the technical sub-committees as well, suggestions made by them were taken on board.

The ESC is, in turn, a sub-committee of the Pakistan Environmental Protection Council, the highest environmental policy-making body, chaired by the prime minister or his/her direct nominee. This kind of access to the political establishment can ensure quick and binding decisions. Unfortunately, however, while the ESC remained extremely active in developing proposals and recommendations to its parent body, the frequency and regularity of PEPC meetings were adversely affected by changing political tides in Pakistan. The life expectancy of a government remains unpredictable, national priorities have constantly changed, and there has been a perpetual state of economic crisis in the country. This has especially been the case after the imposition of sanctions following the nuclear tests in May 1998, which prevented the civil government, and subsequently the military government, from fulfilling their environmental obligations and commitments.

The following recommendations for future action emerge from the above discussion. Some may need the backing of specific technical or financial support, while others simply suggest sustaining the positive momentum achieved so far.

- *Institutionalizing government–business dialogue* In order to maintain the momentum and level of trust established in the consultative process, it is necessary to institutionalize the existing arrangements for policy dialogue. A permanent platform is needed to allow for information exchange, networking and policy dialogue. Ideally, such fora should exist at both provincial and national levels.
- *Need for public pressure* One of the complementary forces that could help keep attention on the industrial pollution abatement programme is public pressure on the government and private sector. This force, however, has not yet been sufficiently tapped. The partnership in this effort must now grow, therefore, to include advocacy groups, media, consumers and other environmental activists.
- *Enforcement of existing and agreed pollution prevention regulations* It is fortunate that a comprehensive set of environmental regulations is now in place. The regulatory agencies should take advantage of the fact that the detailed proposals for implementation of NEQS have been jointly developed with the private sector. The government must now enhance its commitment and capacity to enforce these.
- *Technical assistance* In addition to the technical assistance required by regulatory agencies (including NGOs), an institution must be identified that can address industry's needs for technical assistance and information exchange, undertake research and development on industrial waste management, develop programmes on economic

incentives, and provide linkages to national and international organizations to facilitate the transfer of clean technology.

- *Elimination of perverse subsidies* The government does not systematically take into consideration environmental concerns in its national development planning cycles. Research is therefore required to identify and eliminate existing policies that promote economically inefficient and environmentally unsound practices.
- *Green cash required* Desperately needed green credit lines are unlikely to be established by the government or commercial banks any time soon. International donors, including private sector lending agencies, must be encouraged to stimulate environmentally sustainable industrial development in Pakistan, with the provision of soft credit for environmental projects. The only way this can realistically happen is if a suitable financial and political climate is restored to the country.

In conclusion, there is reason for cautious optimism. Opportunities exist, albeit as tough challenges in the face of today's political and economic realities, to make a permanent transition to sustainable industrial development in Pakistan. If the momentum generated thus far by the negotiation process can be sustained, there is a strong chance that the pollution charge regime will become an institutionalized mechanism for industrial pollution control. The beauty of the programme, in addition to its simplicity and transparent formulation, is that it is a completely indigenous effort without any 'donor pressure'. This fact, in and of itself, also accounts for its sustainability.

The pilot phase of the self-monitoring programme is expected to take three to four more months, when the government will have collected some information on the status of compliance or non-compliance with NEQS. These data, as agreed, will not be used against the participating units during this period for imposing penalties. After this phase, the self-monitoring and reporting programme will be extended to the entire country on a mandatory basis. In the meantime, the government will be required to finalize and notify all relevant rules and regulations to reinforce this system. During the implementation process, it is also expected that the government will identify the highly polluting groups of industries and prioritize and focus regulatory efforts on them. A full-fledged support structure at the national level is also expected to grow side by side with these activities in order to provide analytical and technical services to industry for monitoring and pollution control.

The pollution charge system, which, as earlier stated, is the key element for industrial pollution control, is at an implementing crossroads. All the necessary details, such as procedures and schedules, have already been agreed. In fact, industry representatives even took the bold initiative of suggesting the base rate of a pollution unit and the schedule of increase over a five-year period. The structure is sensible because it initially represents a slap on the wrist, but by the sixth year 'draws blood' so that it would be cheaper to adopt mitigation measures than pay the fine.

There are two issues that may require renewed consultations between government and industry. First, the institutional arrangements for the collection and use of funds have not yet been agreed to. There exists a conceptual difference between rules framed by government for Provincial Sustainable Development Funds (PSDFs), to be created under the existing law, and the operational mechanisms of Provincial Environmental Trust Funds (PETFs), proposed by the private sector. The government negotiators assured the private sector during ESC proceedings that the existing rules could be revised, without any amendment in law, to address all their concerns. Second, the sub-committee on pollution charges recommended to the ESC to change the term 'pollution charge' to 'environmental improvement contribution'. The private sector has its own public relations reasons for pushing this. The ESC could not decide on this issue since the change would require an amendment in the existing Act. For now it seems that the government is sticking to the term 'pollution charge' and that business is going along with this.

When the self-monitoring and pollution charge systems are implemented, industry will require technical assistance as well as soft loans for installing pollution-control equipment and for initiating environmental management activities. Without these, nothing much is expected to change on the ground. Regular and institutionalized contact between government, industry and service providers will be necessary during this phase. There also needs to be a regular assessment of genuine limitations of different industrial sectors in implementing the existing laws and the creation of institutions to look after specific issues and provide policy advice. Ultimately, transparency and across-the-board implementation is going to be the deciding factor in the success or failure of the entire programme.

Notes

1. Comments on earlier drafts are gratefully acknowledged from Dr Mahmood Ahmad Khwaja, Mahmood Ahmed, Zaffar A. Khan and Shahrukh Rafi Khan. The latter's extensive work as editor in updating and adding to the chapter is also duly acknowledged.

2. The re-establishment of the National Environment Standards Implementation Committee by the Ministry of Environment in late 2001, under the chairmanship of environment lawyer Dr Pervez Hussan, is meant to ensure that implementation of the National Environment Quality Standards picks up momentum.

3. A pollution charge is a fee or tax on the amount of pollution in excess of levels allowed by the NEQS. The aim of the charge is to discourage environmentally damaging activities and/or strengthen incentives to reduce waste and pollution, while at the same time generating revenue that may be earmarked for environmental protection.

4. Article 11(1) of the Environmental Protection Act 1997 provides the legal basis for NEQS compliance: 'Subject to the provisions of this Act and the rules and regulations made thereunder no person shall discharge or emit or allow the discharge or emission of any effluent or waste or air pollutant or noise in an amount, concentration or level which is in excess of the National Environmental Quality Standards or, where applicable, the standards established under sub-clause (i) of clause (g) of sub-section (1) of section 6.'

5. The NEQS consist of 32 liquid and 16 gaseous parameters in addition to limits on noise pollution.

6. It has been agreed through consensus that initially only ten liquid and seven gaseous NEQS parameters will be charged for. The list was arrived at in view of the following considerations: (i) to keep the system simple and cost-effective and (ii) the quantity of pollutant defined as one pollution unit reflects the relative toxicity of the pollutant, and consequently the extent of damage to the environment and to human/worker's health. Other NEQS parameters will be phased in on an agreed schedule.

7. *Guidelines for Determination of a Pollution Charge for Industry*, March 1998.

8. The proposed pollution charge of Rs. 50 per pollution unit will be achieved by charging 10 per cent in year one and escalating to 80 per cent of the base rate in year five.

9. Although renaming the charge was endorsed by the ESC, its official adoption would require an amendment to EPA 1997.

10. 'The Provincial Sustainable Development Fund shall be derived from the following sources:

(a) grants made or loans advanced by the Federal Government or the Provincial Governments;

(b) aid and assistance, grants, advances, donations and other non-obligatory funds received from foreign Governments, national or international agencies,

(c) and non-Governmental organizations;

contributions from private organizations, and other persons'. Article 9(2).

11. The SDPI's support for the Environmental Standards Committee has been made possible through this programme.

PART III

*Emerging Issues for the South
on the Trade and Environment
Interface*

CHAPTER 6

The Kyoto Protocol's Clean Development Mechanism: What is it, and what's in it for the South?

Aaron Cosbey and Victoria Kellett[1]

§ Climate change is a global threat – to the earth's environment, the well-being of its people and the strength of its economies. The planet's climate is changing due to the build-up of human-made greenhouse gases (GHGs) such as carbon dioxide (CO_2), methane (CH_4) and nitrous oxide (N_2O). Without the naturally occurring greenhouse effect, which traps heat from the sun near the Earth's surface, the planet's climate would be too cold to support life. Since the Industrial Revolution began in the eighteenth century, however, atmospheric concentrations of GHGs have risen significantly as a result of industrial and land use practices, which were in turn fed by population growth, world wars, rising affluence, and an increasing dependence on fossil fuels that has come to seem almost inseparable from economic growth. Scientists widely predict that these rising concentrations of GHGs will heat up the planet. Scenarios provided by the Intergovernmental Panel on Climate Change (IPCC), an international body of experts, estimate an increase in the mean global temperature of between 1 and 3.5 degrees celsius by 2100. This seemingly small change is nevertheless greater than the planet has seen over the past 10,000 years. It masks significant regional disparities; indeed, those regions that are most vulnerable may well be those least able to cope with rapid change.

Many predicted impacts of climate change, such as coral bleaching, thawing permafrost and the spread of tropical diseases, are already being observed. Over the next 50 to 100 years, other critical effects of climate change are expected to include ocean warming and sea level rise, flooding in low-lying areas and small islands, changing ocean currents, intensification of weather trends and more frequent extreme weather events, and changing habitats and environments as local ecosystems struggle to keep up with the pace of change. Human health will be affected due to the spread of diseases to non-immune populations, an increase in the quantity and range of environmental allergens, and more frequent heatwaves, among other effects. Food security and economic activity may be significantly affected as climate change is likely to happen not smoothly and gradually, but in a series of unpredictable lurches that make adaptation difficult for societies and ecosystems alike.

Climate change has been on the international political agenda since the mid-1980s, when the IPCC was created. It has only been since the late 1990s, however, that the international community has started seriously to tackle the need to mitigate human-made GHG emissions.

Arguably, the most controversial element of the budding international regime to combat climate change is the Clean Development Mechanism (CDM). Cooked up from various proposals in the last hours of negotiating the Kyoto Protocol (see below), the CDM allows developed countries to earn credit for certain types of investments in developing countries. Proponents insist that, properly implemented, the CDM will produce wins for the environment, for industry, and for poor and rich countries alike. Opponents claim that the CDM is another example of 'environmental colonialism' – rich countries avoiding real changes at home by exploiting cheaper options in poorer countries.

This chapter proposes that the reality is likely to be somewhere in between those views. It argues that the CDM offers developing countries real opportunities for improving their domestic environments and pursuing economic and social objectives while contributing to a reduction in global GHG emissions. But it cautions that fully exploiting these opportunities will require that those countries take strategic action, both in terms of understanding how potential CDM investments relate to public policy objectives, and in terms of making policies that attract 'desirable' investments. And it also cautions that many of the fundamental elements of the CDM remain to be negotiated.

The debates surveyed in this chapter are fundamentally about trade, investment and environment. The CDM is an economic instrument to foster investment for environmental improvement and it will use trade in credits earned as a key mechanism for its viability. Thus, while it is in the context of an environmental agreement, the CDM and, in fact, the Kyoto Protocol are in essence agreements about economies and how they should be structured.

What is the Clean Development Mechanism?

How we got there: building the climate change regime At the 1992 Earth Summit in Rio de Janeiro, about 150 countries agreed to the United Nations Framework Convention on Climate Change (FCCC). In doing so, they acknowledged the need for preventive, precautionary action to slow down climate change by limiting emissions of human-made greenhouse gases.

The FCCC divided countries, or parties to the convention, into two groups. Parties listed in Annex I of the Convention are the industrialized countries and those with economies in transition. Non-Annex I countries are chiefly developing countries, with much lower per capita GHG emissions. The ultimate goal of the FCCC is to achieve stabilization of GHG concentrations in the atmosphere 'at a level that would prevent dangerous anthropogenic interference with the climate system' (FCCC Article 2). To this end, Annex I parties agreed in Rio to a non-binding target to limit their GHG emissions to 1990 levels by 2000.

On 21 March 1994, the FCCC came into force.[2] However, it was quickly realized not only that the non-binding FCCC targets were insufficient for meeting the goal of the Convention, but also that most Annex I parties would not even come close to achieving their targets. In 1995 in Berlin, the first Conference of the Parties (COP-1) to the FCCC established the 'Berlin Mandate' to begin negotiations towards a protocol that would contain tougher, legally binding emissions reduction targets.

In December 1997 at COP-3 in Kyoto, Japan, those negotiations culminated in the Kyoto Protocol to the FCCC. Under Article 3.1 of the Kyoto Protocol, Annex I parties agreed to Quantified Emissions Limitation and Reduction Objectives (QELROs), or targets, to reduce their overall emissions of six GHGs to 5.2 per cent below 1990 levels by the period 2008 to 2012.[3] Through the negotiations, this overall target was translated into individual targets for each Annex I Party, which are specified in Annex B to the Kyoto Protocol.

Under the Protocol, Annex I parties agreed to implement policies and measures domestically to reduce their GHG emissions. However, they also negotiated three mechanisms to increase the flexibility available to them in meeting their targets. International Emissions Trading (Article 17) refers to the buying and selling of parts of a party's assigned amount of GHG emissions (termed Assigned Amount Units) between Annex I parties. Joint Implementation (Article 6) is a project-based mechanism by which an Annex I party investor earns credit (called Emissions Reduction Units) for investments made in other Annex I countries that reduce emissions in that country. The CDM (Article 12) is also a project-based mechanism. Under the CDM an Annex I investor can earn credits (called Certified Emissions Reductions or CERs) for investments made in non-Annex I ('host') countries that contribute to sustainable development in the host country and also reduce GHG emissions against an agreed-upon baseline.

For the Kyoto Protocol to enter into force it must be ratified by at least 55 countries, including Annex I countries that represented a minimum of 55 per cent of Annex I emissions in 1990.[4] As of 27 November 2000, the Protocol had been ratified by 31 parties, all of them non-Annex I. Many Annex I and developing country parties are delaying decisions on ratification until there is greater certainty regarding how the Protocol will be implemented. This will depend on the outcome of negotiations to clarify and elaborate key provisions of the Protocol, including mechanisms, compliance, and the inclusion of carbon 'sinks' – land uses and practices, such as reforestation, that absorb carbon from the atmosphere.

CDM: the 'Kyoto Surprise' Among the many elements of the Kyoto Protocol requiring further definition and elaboration, the CDM is one of the most prominent and complex. It is often referred to as the 'Kyoto Surprise', emerging as it did at the last minute from the convergence of three separate tracks.

The first was the pilot phase, begun in 1995, of what was originally known as 'joint implementation'. At that time joint implementation, as advocated by many Annex I countries, implied an Annex I country investing in a developing (or other) country in return for emissions reduction credits. Developing countries did not agree to the proposal, but finally agreed to a voluntary, non-creditable pilot phase on 'activities implemented jointly' (AIJ). The AIJ pilot phase would eventually be reviewed, lessons would be learned, and joint implementation – for credit – would become formalized. By the time of the Kyoto negotiations in 1997, however, key developing countries vocally opposed joint implementation; they advocated continuing the pilot phase of AIJ rather than instituting a fully creditable joint implementation system.

The second factor that led to the CDM was the US Senate's Byrd-Hagel Resolution of 21 July 1997. The Senate opposed the signing of any agreement that would mandate reduction targets in the USA while exempting developing countries from assuming any limits on their own emissions. Realizing the political and diplomatic futility of calling for developing country targets in Kyoto, and in the interests of reaching agreement, the Clinton administration interpreted the Resolution loosely as a call for 'meaningful participation' in the Protocol by developing countries. The USA and allies strove to achieve this by seeking a mechanism that would both involve developing countries in international efforts to reduce emissions, and lower the cost of compliance for the USA.

The third track was a Brazilian proposal, put forward in 1997, to establish a 'Clean Development Fund' that would use proceeds from Annex I non-compliance penalties to finance emissions reduction and adaptation projects in developing countries. This proposal enjoyed widespread support from developing countries. In Kyoto, as discussions stalled on joint implementation, a US-led group of Annex I countries seized upon the proposal. They negotiated on the aspects relating to compliance, and focused on the potential role of a 'Clean Development Mechanism' in allowing Annex I investors to earn credits or CERs on the basis of projects in developing countries, where emissions reductions are cheap. This would expand the range of options available to Annex I parties in meeting their targets, greatly reduce the cost per ton of emissions reductions, and involve non-Annex I countries in the process. Agreement was finally reached during the last hours of COP-3, as Annex I countries, carefully avoiding framing the CDM merely as 'joint implementation with credit', traded agreement on targets for agreement on the CDM.

Article 12 of the Kyoto Protocol specifies that the purpose of the CDM is to assist non-Annex I parties 'in achieving sustainable development and in contributing to the ultimate objective of the Convention, and to assist Parties included in Annex I in achieving compliance' with their emissions reduction targets. The CDM will be overseen by an executive board, which will be financed by a share of the proceeds from CDM projects. Emissions reductions achieved under the CDM will be certified by operational entities and, in principle, could have started as early as 2000. Certified Emission Reduction (CER) credits can be banked between 2000 and 2012 for use during the first commitment period of the Kyoto Protocol in 2008 to 2012. This gives the CDM an advantage over international emissions trading and (Annex I) joint implementation, which begin in 2008.

Key Issues to be Resolved in the CDM

While CDM activities were technically creditable from 2000, several critical issues arising from the Kyoto Protocol needed to be resolved before anyone had a clear idea of how the mechanism would actually work. Some of these were technical issues related exclusively to the CDM; others were more general issues that would nevertheless have a significant influence on the size and shape of the mechanism. At COP-4 in 1998, parties gave themselves until COP-6 at The Hague in November 2000 to agree on the outstanding issues. But it was an indication of the complexity of the issues, and the breadth of their political and economic implications, that no agreement was reached at The Hague. Not only are the key issues still to be resolved, but it is even clearer that it will take several years of 'learning by doing' before the CDM is fully defined and understood.

Project cycle and project eligibility At its simplest, a CDM project cycle would involve at least the following four steps: 1) project design, development and financing; 2) project validation and registration with the CDM executive board; 3) project implementation and monitoring; and finally 4) verification and certification of emissions reductions. 'Operational entities', independent third parties that would be accredited by the Executive Board, would perform such functions as validation, verification and certification.

Nevertheless, within these parameters there remain a number of uncertainties. For example, at what stage does one ensure, and who is responsible for ensuring, that a project will contribute to sustainable development in the host country? How could this be measured? Mindful of the long history of unsustainable projects financed by development agencies and approved by host countries, some observers support the establishment of global principles or guidelines for hosts and investors. This is unlikely to happen, at least in the early years of the CDM. Most developing countries maintain that, as sovereign states, they alone may define their national sustainable development priorities and decide whether a project meets this requirement as part of their project approval process. This could lead to differences in the types of project accepted in different countries, which might create an inappropriate sense of competition between sovereign states to receive CDM financial flows into their economies. One way of ensuring that CDM projects both contribute to sustainable development and reduce GHG emissions is to define specific types of projects that may be eligible or ineligible to earn CERs under the mechanism. This issue has been hotly debated. Many parties, for example the European Union and many environmental non-governmental organizations, favour a 'positive list' of types of projects (such as solar or wind energy) that are eligible for CDM. Such a positive list could be *exclusive*, meaning only project categories included on the list are eligible, or *indicative*, meaning projects on the list might be considered automatically eligible or could be 'fast-tracked' through certain stages of the project cycle, whereas projects not on the list would have to demonstrate eligibility. Others support a 'negative list' of project categories that are explicitly ineligible (nuclear energy is the most notable example, but many also wish to exclude large hydro-electricity projects). Still other parties, such as Canada and the USA, oppose any kind of project list in the interests of maximizing the flexibility for earning CERs.

One of the most controversial questions facing the CDM is whether projects that aim to absorb carbon from the atmosphere will be eligible to earn CERs. It is argued that such carbon 'sink' projects, such as reforestation, preservation of threatened forests or enhancement of a watershed, might offer phenomenal collateral benefits to some developing countries, thus easily qualifying as contributing to sustainable development in the host country. For example, Columbia argues that the reforesting of a watershed in the Andes will stabilize water supply in downstream cities, support biodiversity, prevent erosion and help stabilize local climate.[5] On the one hand, many developing countries, especially in Africa and Latin America, feel that allowing sink projects may be their only chance of attracting CDM investment, as some of these countries do not have extensive GHG-emitting industry. Without sinks, they argue, cost-effective emissions reductions opportunities would be clustered in the few industrialized countries, such as China, India and Brazil, that currently attract the bulk of foreign direct investment.

On the other hand, many countries argue that while it is imperative to protect forests and other carbon sinks for many reasons, this should be done outside the framework of the CDM. Their primary concern is that sink projects are a loophole that will allow Annex I parties to avoid or delay implementation of needed emissions reductions in their own sectors that contribute the most to the problem, mainly transportation and industry. Second, they worry about so-called 'leakage' – if a project restricts people's access to land, timber or other forest resources, it is very likely that they will simply exploit similar resources elsewhere. The result is that emissions are simply displaced rather than reduced. The third concern relates to the impermanence of sink projects. There is no guarantee, for example, that a forest fire might not reverse all the benefits of the project built up over time, and yet the investor will still be left with the CERs earned. In more general terms, there are major scientific and methodological uncertainties relating to predicting and quantifying carbon sequestration, and thus quantifying the emissions reductions from a project might be extremely difficult. Finally, some countries believe that allowing investors to pay for, and get credit for, certain (limitations on) land-use practices in the host country would constitute an unacceptable infringement on their territorial sovereignty.

With powerful arguments on both sides of the issue, it remains to be seen whether sink projects will be included in the CDM.

Additionality and baselines No matter what type of projects are ultimately included under the CDM, each project will be eligible only if it meets the requirements of two important provisions of the Kyoto Protocol. Article 12.5 (c) specifies that emissions reductions must be 'additional' to any that would occur in the absence of the project, and Article 12.5 (b) states that emissions reductions should result in 'real, measurable, and long-term benefits related to the mitigation of climate change'. The two issues arising from these provisions are known as additionality and baselines.

Since developing countries do not have emissions reduction targets, there must be a way of determining the amount of emissions reduction achieved by a certain project. It is therefore necessary to establish a baseline, or business-as-usual scenario, which quantifies the emissions that would have occurred in the absence of the project. Baselines can be project-specific or standardized.[6] For example, consider a windpower development in Costa Rica that is displacing part of an oil-fired power source. A project-specific baseline would attempt to quantify the specific emissions from the oil-fired source. A standardized baseline may state that all on-grid projects are assumed to displace the average emissions of the entire electricity system, including all hydro, oil, and other fossil fuel sources.

On the one hand, proponents of the project-specific approach stress that baselines should be as accurate and detailed as possible to avoid the risk of an investor overstating the emissions reductions achieved. Project boundaries must also be clearly delineated to minimize leakage, which occurs when emissions that were avoided through one project simply occur elsewhere or at a later date. Project-specific baselines are particularly appropriate for those projects with a large magnitude of emissions reductions. On the other hand, the transaction costs associated with producing a baseline scenario for each project may deter many potential investors, particularly with smaller projects. Some countries therefore advocate standardized baselines, which are based on an emissions performance standard for a particular sector in a particular geographic area. This would lower the transaction cost for the investor. However, it may take some time – and some debate – to establish standardized baselines for each sector, and the host country may well be the one footing the bill. For this reason, one approach might be to begin with project-by-project baselines and gradually move to standardized baselines. The question then arises as to whether baselines should be static or dynamic. The former offers more certainty to the investor because it is not changed as new information becomes available, but sacrifices accuracy. Dynamic baselines, in contrast, would be regularly adjusted over the lifetime of the project to take account of macro-economic and policy changes, but would result in less certainty for the investor as to the quantity of emissions reductions to be earned.

The notion of additionality has been interpreted in different ways. The various definitions can be roughly grouped into two categories: environmental (or emissions) additionality and financial additionality. Environmental or emissions additionality clearly refers to Articles 12.5 (b) and (c) of the Protocol, and can be demonstrated by showing that project emissions fall below the baseline. If a project is financially additional, it means that it would not have taken place (the funds would not have been invested in the project) in the absence of the CDM. The simplest form of financial additionality addresses the concern of many countries that funds might be diverted from official development assistance towards CDM projects.[7] Developing countries want the CDM to be a source of new funds, not a re-routing of existing aid funds. However, what if the reductions would have taken place anyway as a result of host government policy or through foreign direct investment? Would they then be additional? It has been argued by some that this form of additionality might be impossible to measure, owing to the difficulty inherent in establishing the business-as-usual scenario, which would involve predicting government policies, investment decisions and their outcomes. Still other forms of additionality are regulatory and technological additionality. A project would be regulatory-additional if it clearly exceeded a regulatory standard or met a regulatory standard that is not normally enforced. Technological additionality refers to whether a project uses the best available technology that would not otherwise be available to the host country.[8] The parameters for additionality, as for baselines, will influence the size and shape of the CDM. In the end, it will come down to a trade-off between the environmental integrity of the mechanism (and the Protocol) and the need to minimize transaction costs to ensure the CDM's attractiveness for investors.

Fungibility and developing country market participation The three Kyoto mechanisms will generate three types of tradable commodities: CERs, emissions reduction units (from joint implementation) and assigned amount units (from international emissions trading). It is likely that these various units will be expressed in the same way, probably in tonnes of CO_2 equivalent. On the one hand, many Annex I parties and some developing countries see no reason that credits generated by the three mechanisms should not be fungible or convertible – that is, they will have the same value, command the same price, and be equally valid for compliance purposes – and thus traded on a secondary market. This could facilitate projects financed within developing countries and then trading the CERs on a secondary market. Many developing countries, on the other hand, insist that CERs should not be treated as a commodity, but should be used exclusively by the investing (Annex I) country for the purposes of meeting its Kyoto target. The fundamentally different approaches of these two camps is largely a reflection of their view of the role of the market in achieving environmental goals. Many developing countries, such as China and India, believe that Annex I countries have a moral obligation to change their lifestyles and consumption patterns (particularly given their overwhelming share of responsibility for historical emissions) and fear that the mechanisms, especially the CDM, are simply a way of avoiding this obligation.

It should be remembered that developing countries, without emissions targets, have few means of participating in international emissions markets. Non-fungibility of CERs could pose a problem for host countries who wish to negotiate some kind of benefit-sharing arrangement, whereby the CERs resulting from the project are split between the host and investor. If the host country could not sell the CERs, what then? Should it bank them for use in the future? A non-Annex I party might feel that banking of CERs for eventual use brings it several steps closer to the decidedly unwelcome prospect of binding emissions reduction targets.

CDM architecture and equitable geographic distribution An even stronger argument for fungibility relates to the fundamental question of just who will carry out stage one of a CDM project, i.e. project design, development and financing. Several possible models have been advanced relating to implementation of the CDM. Each model has various pros and cons depending on the interests of the potential host country.

It was initially assumed that an Annex I entity would carry out stage one of the project cycle, even if host country entities played a large role in project implementation and monitoring (stage three).[9] After all, it is the Annex I entity that has a target and therefore needs the CERs. This basic model is known as the *bilateral* CDM. Experience with AIJ has shown that this model can entail high transaction costs due to the need to identify, negotiate and finance a project on an individual basis. A bilateral CDM would thus favour large projects in a few countries that bigger investors already know well, due to their experience with foreign direct investment. Least developed countries (LDCs), which generally have low per capita emissions, may be left out owing not only to their low capacity to absorb investment but also to a shortage of emissions mitigation opportunities.

An alternative vision of the CDM is the *multilateral* approach, which is comparable to a mutual fund. Under this model, investors would channel resources into the fund, which would be responsible for designing, developing and financing the CDM project. Investors would then receive CERs proportional to their initial investment. The advantages of this approach may be reduced risk for investors and reduced transaction costs due to pooling of resources and expertise. It has also been noted that multilateral funds might be managed by entities such as regional development banks, whose mandates may be more sympathetic to developing countries' interests. In comparison to the bilateral model, this might allow host countries greater power when negotiating CDM projects, leading to more equitable outcomes; it might also broaden the range of countries that would host CDM projects. Thus far, the most prominent example of a multilateral approach is the World Bank's Prototype Carbon Fund, established in 1999. It is capitalized to a maximum of US$150 million by contributions both from Annex I governments and from companies. As of June 2000, 18 developing countries had signed on, including seven LDCs.[10]

A further possible model is the *unilateral* model, characterized by the absence of an Annex I party investor. Under this model, the host country would develop, finance and implement CDM projects, taking on all the risks themselves, but also retaining the resulting CERs. This approach assumes, of course, that the host country will be able to sell, bank or otherwise use the CERs. Proponents of this model are generally countries that have the capacity to 'go it alone' from project conception to completion, with the exception, of course, of the required third-party stages of validation, registration, verification and certification. By definition, unilateral projects are most likely to be compatible with national sustainable development priorities. Some potential host countries, moreover, have difficulty attracting foreign investment due to high country risk and transaction costs; they believe the unilateral approach is their best chance at participation in the CDM. However, a great many countries, particularly LDCs, do not have the capacity to implement unilateral CDM projects, and thus would be left out of the CDM under this model. A variation on the unilateral model is the *South–South* model, whereby a project implemented in the host country would be developed and financed by an investor from another developing country. Like the unilateral approach, this model assumes that a non-Annex I party will then be entitled to sell or use the CERs generated from the project. This model may be particularly useful among developing country groups with close economic and cultural ties.

It has been argued that a combination of the bilateral, unilateral and multilateral models would best meet the needs of both the various host countries and of investors, and thus both maximize the flow of investments and CERs and help ensure an equitable geographic distribution of projects by allowing means for participation of LDCs, high-risk countries, low emitters, and others. As the CDM becomes a reality and lessons are learned through experience, one or more models may become more dominant, while others may be discarded.[11]

Adaptation levy and competitiveness of the CDM   A further uncertainty regarding the CDM is the implementation of Article 12.8, which states that a 'share of the proceeds' from CDM projects will be levied to cover administrative costs and to assist vulnerable developing countries in adapting to climate change. This so-called 'adaptation levy' raises a number of questions. For example, what is meant by a 'share of the proceeds?' Does this refer to the number of CERs generated by the project, or to the value of the CERs? Or even to the profits generated? How much of this levy should be allocated to administration fees, and how much to adaptation? What proportion of the proceeds should be levied to allow the maximum flow of funds to vulnerable countries without deterring investors? Should there be a minimum project size before the levy kicks in, so that small-scale projects can be more competitive?

Many potential investors claim that the adaptation levy will increase their transaction costs, and thus possibly decrease the attractiveness and competitiveness of the CDM *vis-à-vis* the other mechanisms. They therefore favour minimizing the levy. For the same rationale but from a very different perspective, some developing countries and non-governmental organizations advocate expanding the scope of the levy to cover international emissions trading and joint implementation, both to avoid putting the CDM at a disadvantage and to maximise revenues for adaptation. Many LDCs and small island states favour exempting projects carried out in their countries from the adaptation levy. Recent negotiations indicate fairly wide support for this argument; since LDCs and small island states are likely to be the primary recipients of adaptation levy revenues, it would not make sense to tax investments in their countries, only to return the funds in another form. Also, projects in these countries are likely to be small in number and size, so the forgone adaptation levy would be minimal.

Others argue that competitiveness may not be such a problem. They note that the CDM already has several advantages over the other mechanisms, namely that it begins in 2000 rather than 2008, that it may supply lower-cost reductions than those effected in Annex I countries, and finally that, unlike international emissions trading and joint implementation, it is not part of a zero-sum game. In other words, if Annex I countries rely heavily on the mechanisms to meet their target, the demand for assigned amount units, emissions reduction units and CERs will escalate. Since the total supply of assigned amount units and emissions reduction units is limited by the collective Annex I target, demand for credits will be channelled to the CDM or to domestic emissions reductions, whichever is economically and politically cheaper for the party trying to comply with its target.

Supplementarity Another issue that has been the subject of protracted negotiations is the extent to which Annex I parties may rely on the Kyoto mechanisms versus reducing GHG emissions domestically. As mentioned above, the Kyoto Protocol calls upon Annex I countries to implement policies and measures at home to reduce their domestic emissions. It was agreed in Kyoto that a party's use of the three Kyoto mechanisms be 'supplemental to domestic actions' to meet its target.[12] This notion of 'supplementarity' embodies the perception of many developing countries, members of the European Union, and many environmental groups that the mechanisms are an environmental loophole in the Protocol that allows Annex I countries to buy their way out of their obligations. The European Union in particular wants to establish a quantitative cap on the use of the mechanisms; various formulae have been proposed for calculating this cap. However, supplementarity is not defined or quantified in the Protocol. Countries such as Australia, Canada, Japan and the USA prefer to leave it that way. While they have stated their intentions to make emissions reductions at home, they view the Kyoto mechanisms (and domestic carbon sinks) as the most economically efficient means of achieving their target. Negotiations since COP-4 have shown this issue to be highly charged politically. Its resolution will affect the size and the nature of the CDM; a stringent or even moderate quantitative limit on the use of the mechanisms might very well limit the extent of CDM investment.

Opportunities and Pitfalls for Developing Countries

Opportunities As we have stressed, the CDM is billed as more than simply a mechanism for reducing the costs of limiting GHG emissions, but also as a promoter of sustainable development. To what extent does the CDM offer the possibility of real sustainable development, i.e. economic, social and environmental benefits in host countries?

Answering this question with precision is difficult, given the uncertainty described above on the fundamental structure and administration of the CDM. But it is possible to make some general predictions as to the types of opportunity the CDM might offer developing country hosts.

Box 6.1 shows the types of option likely, based on the results of a study that analysed CDM opportunities in Brazil, India and China.[13] As this table would seem to indicate, most of the opportunities will be improvements in the efficiency of conventional fuel use, both of industrial boilers and commercial power generators. There are a number of such opportunities in developing countries, where inefficiency in the power generation sector is one of the reasons the CDM is attractive to investors. There are also opportunities for new investments in renewable energy sources, mostly for power generation.

Another study looked at opportunities focused on the energy sector mainly at the household rural and urban levels, where direct benefits for poverty alleviation might be expected.[14] The projects studied were micro/mini hydro plants, biogas digesters, improved cooking stove and solar home systems. These were located in Sri Lanka, Nepal, Zimbabwe and Kenya.

There may also be opportunities in improving the efficiency of transmission and distribution of electricity, and in improving the efficiency of the transport sector. If forests are allowed as sinks in the Protocol, there is huge potential for some countries to invest in land-use change, forest preservation, afforestation or sustainable forest management. Finally, there is potential for investments in the rural sector: for example, through projects to reduce methane emissions from rice paddies and cattle.

A wide range of project types, then, might be expected to materialize under a working CDM mechanism, with the situation varying considerably from country to country. There are, however, certain types of benefit that might be expected to accrue in most project types.

• Financial and technological inputs The most obvious benefit is the flow of new investment, and the new technology it brings. Efficiency-improving investment lowers the costs of production for firms, and the costs of their products for consumers.
• Capacity-building New investment and new technologies do not exist in a vacuum, but are used by firms in the host country. The new techniques that accompany them, and the knowledge of how

they work, will eventually diffuse to other domestic actors, having the effect of building capacity within the overall economy. Involvement in the CDM not only contributes to technological capacity, but also builds up the host country's experience in doing business with foreign investors and integrating into the global marketplace.

- Local environmental and health benefits Projects could lead to improvement in local air quality, reduced water pollution, elimination of toxic wastes, and/or other benefits. These are usually significant. In the case of increased efficiency of coal-fired electricity generation or the replacement of them with other types of technologies, for

---

Box 6.1 Selected Abatement Opportunities in Brazil, India and China

Conventional power generation
- Combined-cycle gas turbines
- Improved coal technologies

Fuel switching
- Recovery and use of coalbed methane
- Electricity cogeneration from chemical plants
- Fuelwood gasification with pulp residues
- Bagasse-based electricity cogeneration

Industrial applications
- Wide range of efficiency improvements possible in boilers, motors and other equipment
- Modern, energy-saving processes in cement, iron and steel industries

Use of renewables
- Extending biomass fuel sources
- Wind energy
- Solar thermal and solar photovoltaic applications
- Small-scale hydropower
- Wind pumps for irrigation

Forestry options
- Silviculture plantations for pulp, sawlog and charcoal
- Sustainable forest management on private and public lands
- Community woodlots and agroforestry projects

Source: WRI 2000.

example, they include reduced SO_2, NO_x and particulate emissions – a major concern for coal-burning countries, such as China, which suffer acid rain and other air pollution problems.[15] In the case of forest management, they include flood control and preservation of biodiversity.

- Employment With increased investment comes increased employment in the sectors of interest, and alternative energy sources may provide sources of economic development in rural areas not well served by conventional power sources.[16]
- Self-sufficiency In many cases, the projects will result in greater energy self-sufficiency, and reduced import bills for fossil fuels. It is important that developing countries consider this when they are establishing their sustainable development criteria.

These types of non-carbon benefit will accrue to a greater or lesser extent depending on the project and the context. It is argued below that to exploit fully the potential of the CDM, governments will have to deliberately plan to harmonize the results of the mechanism with their existing policy objectives.

Possible pitfalls   One argument often put forward by critics of the CDM is that it will allow Northern countries to harvest the low-hanging fruit from the South – the cheap and easy emissions reduction options – and gain credits for doing so. If and when the developing countries eventually accept binding commitments to act, the argument goes, they will be left with only the most expensive options.[17]

From an environmental perspective this is a poor argument, akin to recommending that an athlete deliberately perform poorly so that he or she can more easily show great improvement later. The problem is that if we wait for the time when developing countries make binding commitments and begin to harvest those cheap opportunities, the interim involves continued environmental degradation and its associated impacts on human health and well-being.

But the argument is not straightforward from an economic perspective either. For one thing, it assumes a static basket of technologies and management practices available for improving environmental performance. If we assume a dynamic basket – a far more realistic assumption, particularly given the inevitable investment in innovation in the OECD countries – it is not obvious that the technological improvements available in the future will be more costly than those now viable under the CDM.

Also, most of the investment that would occur under the CDM is efficiency-enhancing or wealth-creating. Deliberately depriving an economy of such investment is not cost-free. The high costs of inefficiency are inevitably visited on domestic consumers, and on firms that lose market share in international competition. Thus the costs of protecting the lower-cost options for meeting CHG targets may be quite high.

Another concern is that the CDM will encourage investment in modernizing the fossil fuel sector, as opposed to encouraging development of alternative energy sources such as wind, solar, small hydro, biomass and geothermal. Developing countries are forecast to double their 1995 energy consumption by 2020 to accommodate projected population and economic growth.[18] If those investments are made in the fossil fuel sector, where modernization and improved efficiency offer some of the easiest CDM credits, some argue that this will perpetuate the reliance on fossil fuels, which is likely to be incompatible with the FCCC objective of stabilizing atmospheric emissions. Such investments are typically sunk into capital that has a life measured in decades, such as coal-fired generating plants.

This is a valid concern. It assumes, of course, that host countries will not implement CDM project guidelines that favour renewable energy projects. In the end, it boils down to a question of the priorities of the host governments: if they are committed to renewable energy, then they could make the CDM support such a commitment. Several developing countries are expected to prefer renewable energy projects over conventional energy. For example, Costa Rica has stated that it intends to phase out all fossil fuels. Also, members of the World Bank's Prototype Carbon Fund, including industrialized and developing countries, are focusing on renewable energy for the majority of projects.

Nevertheless, large bilateral investors in particular may find the cheapest credits in fossil-fuel projects such as clean coal or coal-to-gas conversion, using familiar technologies and readily available fuels. Some potential host countries may find themselves in a 'race to the bottom' relating to project guidelines. To combat this and to encourage the spread of the cleanest possible technologies, many parties and environmental groups want to see a provision in the CDM to fast-track certain projects types, notably renewable energy, making them more attractive to investors. Developing countries, many of which have limited negotiating power in the global economy, may find that this would help them avoid 'locking in' to a fossil fuel future.

While it would obviously be environmentally preferable if host countries were to adopt explicit commitments to renewable energy, even if they did not do so the CDM might be useful in accelerating the adoption of renewable energy technologies. Many of these technologies are technically and economically viable at present, particularly those that wring new production out of existing processes: bagasse co-generation; gasification of wood or pulp residues; and methane recovery from landfills and coal mining. Windpower is one of the fasting growing sources of electricity in the world today. But many renewable energy options remain unimplemented because of high front-end capital costs. An instrument such as the CDM, which is designed precisely to direct foreign investment into such projects, could help overcome this barrier. To a lesser extent this is also true of the more expensive renewable options, such as solar photovoltaic power.

Implications for Developing Countries

If, in spite of the various pitfalls described above, non-Annex I governments decide that there is merit to participating in the CDM, they should do so strategically, aiming to maximize the so-called 'co-benefits'. A four-stage process for doing so is suggested below.

1. Research the opportunities: Governments must first understand in what domestic sectors there exist opportunities for CDM investment, and what rough level of returns on investment are possible in each. Returns on investment in the sense used here include both financial returns to the investor, and other benefits: environmental, economic and social.
2. Elaborate priorities: Given the diversity of opportunities available, and the types of benefit each might create, which types best fit with existing national priorities? By asking this question, governments will discover what types of likely project are most desirable from a domestic viewpoint.
3. Favour the desirable projects: There are a number of ways in which governments can favour the sectors or project types they find to be most desirable. The most obvious is a screen built into the national-level project criteria. National governments will in some fashion have to approve any CDM projects, and they should be able to establish criteria for eligibility. If pursuing this option, governments must be wary of overly restrictive conditions that might deter favourable investments that they might not have considered in their analysis. Any such criteria should have a light touch, as opposed to a heavy hand. Governments should attempt to minimize transaction costs in the desirable sectors. They will be competing with CDM projects in other countries as well as with the other two Kyoto mechanisms, and so will have to work to make their investments more attractive. There

are a number of ways to minimize transaction costs.[19] A national office for potential investors might be established as a clearing-house of information and approvals. The government might calculate baselines for the most desirable sectors, should sector-level baselines eventually be accepted. Bilateral investment treaties with potential investor countries might also be a way of increasing expected rate of return for investors, since they remove some perceived risk. Such treaties, however, should be carefully analysed for sustainable development impacts.[20] Finally, investors will look for countries with the institutional and technical capacity to absorb easily the technology they propose to transfer. Government investment in training and capacity building in the desirable sectors might help in this regard.

4. Supporting measures: Governments should assess whether there exist domestic policies that might discourage investment in particular sectors of interest. For example, if China identifies new coal technologies as a sector of interest, it might consider liberalizing restrictions on foreign investment specifically in that sector. There may also be opportunity to foster investment in desired sectors through new regulations. Tax breaks for renewable energy technologies, for example, or tighter regulations on the local environmental emissions from conventional sources, will make investment in renewables more attractive.

Conclusions

It is hard to predict how the various elements of the Kyoto Protocol will eventually take shape. The key questions centre on the eligibility of projects (how much influence will developing countries effectively have over project type, and to what extent will they have to compete for investments?); tradability of CERs (will developing countries be able to sell them on secondary markets?); supplementarity (will there be caps on the ability of developed countries to use the mechanisms toward their domestic targets?); and architecture (who will be able to initiate and finance the projects under the CDM?). There are also questions about the baselines to be used, and the power of the host country to set defining criteria.

At the end of the day, however, there does seem to be a possibility that the CDM could offer both of the benefits it is supposed to deliver: reduced costs of limiting GHGs, and sustainable development in developing countries. The type of careful analysis proposed above should help ensure both that investment does come, and that, when it does, its impacts meet with domestic and international objectives.

Notes

1. International Institute of Sustainable Development (IISD).

2. As of May 2000, the FCCC had been ratified by 184 countries.

3. In addition to the most common GHGs, CO_2, CH_4 and N_2O, the Kyoto Protocol includes hydrofluorocarbons (HFCs), perfluorocarbons (PFCs) and sulphur hexafluoride (SF_6). Annex A of the Kyoto Protocol lists the GHGs covered as well as the categories of sources and sectors to be included.

4. It is possible in theory for the Protocol to enter into force without ratification by the USA, which accounted for 36 per cent of Annex I emissions in 1990. However, many observers believe that whatever the formula for entry into force, a *de facto* veto will be held by the USA, given its position as the largest single emitter and its political and economic influence.

5. See, for example, Black et al. 2000.

6. At the extreme, a baseline can be national level. Due to the high level of aggregation and the lack of data and measurement capacity in developing countries, however, national level baselines could be highly questionable, dependent as they are on estimates and uncertain economic forecasts. In addition, the most common argument against national level baselines is that they would be equated with an emissions cap. This is politically unacceptable for the majority of developing countries.

7. Most observers believe that donor countries can, and should, use official development assistance for climate change-related capacity-building activities in developing countries, which among other things will improve their capacity to identify and host CDM projects.

8. For a discussion on various types of additionality, see Tata Energy Research Institute and Pembina Institute for Appropriate Development 2000.

9. Recall that independent third parties will be responsible for project validation and registration (stage two) and verification and certification of emissions reductions (stage four).

10. See the World Bank's Prototype Carbon Fund website at http://www.prototypecarbonfund.org.

11. For a thorough discussion of the various models and their implications, see Baumert et al. 2000.

12. This is the wording found in Kyoto Protocol Article 6.1 (d) relating to joint implementation, and Article 17 relating to international emissions trading. Relating to the CDM, Article 12.3 (b) is worded differently but is interpreted identically. It states that Parties may use CERs to 'contribute to compliance with *part of* their quantified emission limitation and reduction commitments' (emphasis added).

13. See World Resources Institute 2000.

14. See Begg et al. 2000.

15. One study estimates the cost of air pollution in China, for which coal accounts for a major share, at 6 per cent of GDP. See World Bank 1997.

16. See World Resources Institute 2000.

17. See Agarwal 2000.

18. See International Energy Agency/Energy Charter Secretariat 1998.

19. Of course, until the parties have settled the questions raised above on CDM architecture and equitable geographic distribution, we can only speculate on what mechanisms might be available to minimize transaction costs.

20. For a critical assessment of the investment provisions in the NAFTA and their environmental implications, see Mann and von Moltke 1999 and also Chapter 7.

Bibliography

Agarwal, Anil (2000), 'Is the Kyoto Protocol a steal?', *Down to Earth*, 15 March.

Baumert, Kevin (1999), 'Understanding additionality', in Walter Reid and José Goldemberg (eds), *Promoting Development While Limiting Greenhouse Gas Emissions: Trends and Baselines*, New York: United Nations Development Programme, pp. 135–46.

Baumert, Kevin, Nancy Kete and Christiana Figueres (2000), *Designing the Clean Development Mechanism to Meet the Needs of a Broad Range of Interests*, World Resources Institute Climate Note, August 2000. Available online at: http://www.wri.org/cdm/cdm-note2.html.

Begg, K. G. et al. (2000), *Initial Evaluation of CDM Type Projects in Developing Countries*, Centre for Environmental Strategy, University of Surrey. Available online at http://www.surrey.ac.uk/CES/ji/cdm-dfid.htm.

Black, A. et al. (2000), *National Strategy Study for Implementation of the CDM in Colombia, Executive Summary*. Santafé de Bogotá: National Strategy Studies, Ministerio del Medio Ambiente, World Bank.

Center for International Environmental Law (1998), 'Designing a legal and institutional framework for the clean development mechanism', */linkages/ journal/*, Vol. 3, No. 4, 26 October 1998. Available online at: http://www.iisd.ca/linkages/journal/ciel.html.

International Energy Agency/Energy Charter Secretariat (1998), *Energy Investment*, joint paper presented to the G-8 Energy Ministerial, Moscow, 1 April 1998.

Intergovernmental Panel on Climate Change (1995), *IPCC Second Assessment Report*, Geneva: IPCC. Available online at: http://www.ipcc.ch/pub/reports.htm.

Intergovernmental Panel on Climate Change, Working Group II (1997), *The Regional Impacts of Climate Change: An Assessment of Vulnerability. Summary for Policymakers*, ed. Robert T. Watson, Marufu C. Zinyowera and Richard H. Moss, World Bank, Zimbabwe Meteorological Services, Battelle Pacific Northwest National Laboratory, published for the Intergovernmental Panel on Climate Change, November 1997. Available online at: http://www.ipcc.ch/pub/sr97.htm.

Mann, Howard and Konrad von Moltke (1999), *NAFTA's Chapter 11 and the Environment – Addressing the Impacts of the Investor-State Process on the Environment*, Winnipeg: IISD. Available online at: http://iisd.ca/trade/chapter11.htm.

Michaelowa, Axel and Michael Dutschke (1999), 'Economic and political aspects of baselines in the CDM context', in Walter Reid and José Goldemberg (eds), *Promoting Development While Limiting Greenhouse Gas Emissions: Trends and Baselines*, New York: United Nations Development Programme, pp. 115–34.

Rønneberg, Espen (1998), 'Sinks and the clean development mechanism', /linkages/journal/ Vol. 3, No. 4, 26 October 1998. Available online at: http://www.iisd.ca/linkages/journal/ronneberg.html.

Sokona, Youba, Stephen Humphreys and Jean-Philippe Thomas (1999), 'The clean development mechanism: what prospects for Africa?', *Environnement et développement du tiers-monde*. Available online at: http://www.enda.sn/energie/cdm2.htm.

Tata Energy Research Institute and Pembina Institute for Appropriate Development (2000), *Negotiating the CDM: A North–South Perspective*. Available online at: http://www.teriin.org/climate/climate.htm.

World Bank (1997), *Clear Water, Blue Skies: China's Environment in the New Century*, Washington, DC: World Bank.

— Prototype Carbon Fund website, http://www.prototypecarbonfund.org.

World Resources Institute (2000), 'Opportunities for financing sustainable development via the clean development mechanism', in D. Austin and P. Faeth (eds), Washington, DC: World Resources Institute. Available online at: www.wri.org/wri/cdm.

United Nations (1997), *The Kyoto Protocol to the Convention on Climate Change*, Bonn: Climate Change Secretariat. Available online at: http://www.unfccc.int/resource/convkp.html.

CHAPTER 7

The Need for an International Investment Regime

Konrad von Moltke[1]

§ The character of international investment has been changing. Twenty years ago, most international investment was undertaken by large multi-national corporations (MNCs), which sought to secure their raw material supplies or which were establishing production or sales units in the early phases of globalization. Foreign investment was an important adjunct of trade rather than an independent economic activity.

It is a mistake to view foreign direct investment (FDI) simply as an adjunct to trade. Capital is a scarce resource, particularly in developing countries. The efficient allocation of capital is critical to the achievement of economic growth and sustainable development. At present, capital is allocated in a fashion that is counter-intuitive, with the largest flows converging on the most developed countries. In principle, capital should be seeking out the highest returns, which should be available where capital is most urgently needed – in the developing world. The para-mount task for an investment regime is to improve the efficiency of the allocation of capital.

There have been encouraging signs. By the end of the century, foreign direct investment was being undertaken by enterprises large and small with a wide range of underlying concerns. The option of investing in another country has become the normal part of strategic growth plans for many enterprises. Indeed, individual investors are now seeking investment opportunities outside their own currency region as a matter of course. Mutual funds make these kinds of investment available to small investors.

Investment flows have increased dramatically. For many developing countries, FDI has become the most important source of capital inflows, overtaking both official development assistance and the funds made available by multilateral development banks. In developing countries, between a third and a half of private corporate investment is undertaken by affiliates of foreign corporations.[2] Investment flows from OECD to non-OECD countries have finally become positive, with large outflows from more developed to less developed markets. Between OECD countries, traditionally the recipients of the largest amounts of foreign investment, the number of enterprises that participate and the range of projects being funded have grown dramatically.

The growing significance of FDI gives greater urgency to long-standing debates about the creation of an international regime for investments. Investment is not an act of nature. It represents a critical economic function of great social significance. It also has major implications for the prospects of achieving greater sustainability. Mature, strong economies generate and sustain significant levels of investment, and policy intervention is necessary primarily to ensure that essential market disciplines are maintained. Policy-makers in mature economies have grown accustomed to 'having their cake and eating it'. Wealth creation has reached a point where many hard choices can be avoided. In the USA alone, tax receipts rose dramatically between 1992 and 1998 as government participated in an extraordinary increase in the value of many forms of investment. The US government could celebrate the end of budget deficits, maintain a huge defence budget, increase spending on some broadly based social programmes, invest in infrastructure, increase the endowment of its universities, and invest abroad – all at the same time and without raising tax rates. Policy-makers in developing countries do not have that luxury. In developing economies, the need for investment, including investment in infrastructure and social development, is overwhelming, but capital is by definition scarce and it is difficult to ensure that investment flows meet a range of policy objectives.

An international investment regime is ultimately about efficiency and fairness. Insecure or severely distorted conditions of investment are factors of risk, which are reflected in expected rates of return. Countries perceived as representing a high risk will be able to attract investment only for projects that offer exceptional rates of return. Many important projects will remain unfunded, and funds that are available will be used in ways that are not as efficient as they could be. The lack of clear rules also creates an incentive for side-payments and corruption, which again exacts an economic penalty from projects and investment flows.

Issues of fairness arise because of unequal power and conflicting goals of the various participants in investment. Investors from countries that are perceived as weaker will fear for the security of their investments in foreign countries. At the same time, some governments are significantly weaker than many major corporations and may find it difficult to impose their legitimate priorities. At the very least, the effort required to defend less secure investments is itself a drag on investment and a source of economic inefficiency.

Investments are mostly private transactions aimed at generating positive rates of return, but they can have far-reaching implications for the welfare of countries, including the prospects of sustainable development, use and protection of natural resources, and the availability of jobs, incomes and economic security. It is the role of government to balance these sometimes conflicting public and private interests, whether by promoting investment, by creating incentives to direct investment to certain activities or regions, or by maintaining a system of taxes and fees that contribute to public policy goals.

Among the numerous measures confirming the recent growth of FDI, those most likely to identify the interests at stake for various countries concern the 'stock' of FDI relative to GDP and the ratio of trade to GDP.

Small countries with a large involvement in foreign investment – the Netherlands, Belgium and Switzerland in particular – and medium-sized countries whose prosperity depends on international trade and investment – France, the United Kingdom and Canada – all have an urgent interest in protecting their investors. The USA has an interest in an investment regime primarily because the total amount of investment by its citizens is large, even though it is relatively much smaller than that of most other OECD countries.

Those concerned with sustainable development have a particular interest in an investment regime. Many current economic activities, in developed and developing countries alike, are known to be unsustainable. Frequently, alternatives are available, but they require investment. In other words, without investment there is no hope of achieving sustainability, for the simple reason that many current structures are known to be unsustainable. With such an urgent need for investment, the move towards sustainability requires that scarce resources are used efficiently – and that the imperatives of sustainability are respected in the investment process. Indeed, it can be argued that an investment regime that does not actually promote sustainable development represents an important step back from the widely endorsed principles of sustainable development.

In principle, developing countries should welcome an international investment agreement that renders risks more calculable and consequently can reduce the rates of return that must be achieved for investment in developing countries. The need for investment is greatest where current investment is lowest, essentially in developing countries. One measure of the success of an international investment agreement must be whether it encourages more vigorous and more evenly distributed investment in developing countries. Despite these fairly clear reasons for developing an international investment regime, this goal has proved to be surprisingly elusive.

Investment is essential to the achievement of sustainable development. Much existing infrastructure and many existing practices are known to be unsustainable. Investment is needed to replace them. From the perspective of sustainability, it is essential that the scarce resource capital be allocated as efficiently as possible and in a manner consistent with sustainable development. In particular, capital should be flowing to less developed countries and regions – the current pattern of investment represents a form of market failure, which could be corrected by an appropriate international investment regime.

This chapter reviews the long debate about an international investment regime. It considers existing multilateral, regional and bilateral investment agreements and identifies the need for a broad multilateral agreement. It also reviews the two major streams of international debate in this area, one focusing on investor rights and the other on investor obligations, and argues that a multilateral investment regime must address both classes of issue if it is to succeed.

Drawing on experience with international environmental regimes, the chapter identifies the structurally determining characteristics of an international investment regime. It argues that investment occurs over a long time frame and that the relationship between investor and host country is notably different than that in trade in goods or services. Investors acquire rights in the host country, and with those rights come obligations.

The time frame of investment, and the legal rights of investors in host countries, require a dynamic regime. In particular, the principles underlying the process of trade liberalization – most favoured nation and national treatment – and the WTO dispute settlement system are inappropriate for an investment regime. This suggests strongly that an international investment regime needs to be constructed outside existing international organizations, possibly beginning with a framework convention followed by a series of protocols addressing specific issues.

Precursors of an International Investment Regime

There are a surprising number of agreements that can serve as precursors for an international investment regime including both bilateral and multilateral agreements. In building on these precedents, however, it is particularly important to keep in mind that measures that are formally identical can produce dramatically different results, depending on the institutional context.

Bilateral investment agreements The number of bilateral investment treaties is remarkable. By one count, over nine hundred such agreements had been concluded by July 1995.[3] Bilateral investment agreements have typically been concluded between an OECD country and a developing country or between developed countries. It is reasonable to assume that the motivation of one party is generally to attract investment, while the other is seeking to gain some additional protection for that investment.

The bilateral agreements are generally concluded on the basis of 'prototype' bilateral investment treaties that have been elaborated by a number of countries. A complete list of these prototypes is not available, but it must be assumed that the instrument that is ultimately used will reflect the relative power and interests of the two parties.[4]

The prototypes elaborated by OECD countries (France, Germany, Switzerland, the United Kingdom and the USA) base the agreement on the principles of most favoured nation and national treatment. The prototypes from Chile and China enunciate a more general principle of 'fair and equitable' treatment, which is subsequently embedded in what might be termed an MFN (most favoured nation) framework. Chile includes a reference to national treatment, whereas China does not. These differences suggest that there remain some significant differences between countries, even at the level of the underlying principles. 'Fair and equitable' treatment is a relative standard, which not only requires interpretation but can clearly involve different treatment in different situations. MFN and national treatment imply an absolute standard, in particular, in light of their significance in the trade regime. In practice, the remaining need for interpretation and adjustment is to be found in the phrase 'under like circumstances', which applies to both MFN and national treatment.[5]

Bilateral investment agreements include dispute settlement provisions in accordance with generally accepted international legal practice. Most of these provisions draw in one way or another on the International Centre for Settlement of Investment Disputes (ICSID).[6] Alternatively they incorporate the United Nations Commission on International Trade Law (UNCITRAL) Arbitration Rules.[7]

The common characteristic of these bilateral agreements is that they do not 'internationalize' the investment process. The agreements seek to utilize existing national law in a framework that is accessible to nationals of the other country. The international regimes that are created are minimalist, lacking individuality or any institutional or organizational capabilities. They generally rely on ICSID for dispute settlement. ICSID in turn remains embedded within the World Bank, sharing facilities and, to a certain extent, personnel.

Multilateral approaches to an international investment agreement It is tempting to view multilateral efforts to establish an international investment regime as little more than extending bilateral agreements to the multilateral level. Such a view misses the essential institutional differences between a regime that draws on existing national law and one that seeks to create new international law. In practice, these are entirely different regimes.

It is possible to distinguish two fundamentally different approaches to developing a multilateral agreement on international investment.[8] A series of initiatives dating back to the origins of the trade regime has sought to define rules governing the treatment of foreign investment by states, generally by the 'host' state in which the investment is made. These initiatives focus on the rights of investors and a dispute settlement procedure to ensure that these rights are respected. An alternative approach, based largely on the United Nations Conference on Trade and Development (UNCTAD) and its United Nations Centre on Transnational Corporations (UNCTC), sought to define the obligations of corporations that invest in foreign countries. Both approaches have attracted strong opposition, in the first instance mainly from developing countries, in the latter from certain OECD countries, and from the USA in particular, leading ultimately to the closing of the UNCTC.

The controversies surrounding these attempts to address the issues relating to international investment have led to a situation where not even the questions have been properly framed. No attempt has thus far been undertaken to draw both approaches together and to develop an agreement that encompasses both the rights and the obligations of foreign investors. The most comprehensive study of the law of international investment concluded:

> Multilateral instruments, whether binding or non-binding, are difficult to construct in the area of foreign investment. A satisfactory code must address both the issue of how foreign investors conduct themselves in host states as well as the treatment of foreign investors by the host states. For historical reasons, the two bodies that have produced recent instruments are identified with different camps in the debate. If a code is to be produced, it is better that the attempt is made by other institutions or a new institution.[9]

Investor rights The Punta del Este Declaration, which defined the negotiating mandate for the Uruguay Round of trade negotiations, stated that 'following an examination of the operation of GATT Articles related to the trade-restrictive and distorting effects of investment measures, negotiations should elaborate, as appropriate, further provisions that may be necessary to avoid such adverse effects on trade'.[10] This mandate was linked tightly to the trade-related aspects of investment, and the resulting Agreement is correspondingly modest in scope.[11] Article 1 specifies that it applies to investment measures related to trade in goods *only* (italics added) and Article 9, which sets up a review of the Agreement no later than five years after its entry into force (2001), states that 'in the course of this review, the Council for Trade in Goods shall consider whether the Agreement should be complemented with provisions on *investment policy and competition policy*' (italics added).

In substantive terms, the TRIMs Agreement extends the protections of Article III (national treatment) and Article XI (quantitative restrictions) of GATT 1994 to trade-related investment measures. In an Annex, it includes five measures that are 'inconsistent with the obligations of national treatment', as an 'illustrative list' (in practice a limitative list). These are largely what are known as 'performance requirements' in the OECD context. The distinction between TRIMs and 'investment policy' as well as the inclusion of 'competition policy' in Article 9 provide important guidelines to the limits of the Agreement and the essential differences between the TRIMs Agreement and a multilateral investment regime.[12] By distinguishing between TRIMs and 'investment policy', the Agreement clearly implies that the two are not the same and may require entirely different international disciplines.

The Uruguay Round also revealed deep and abiding differences of opinion between major participants in the negotiations. The USA and Japan had the most expansive view of 'trade-related' investment measures. The European Union had a somewhat more restricted perspective. The Nordic countries took a position that was even more restrictive. Key developing countries argued that investment measures are legitimate instruments in the context of their economic situation and advocated a case-by-case approach to TRIMs. Over the past few years, however, attitudes of some developing countries to investment have changed. In particular, the countries that have been most successful in attracting foreign direct investment have shown a much greater willingness to embrace a regime based on the principles espoused by OECD countries.

The World Bank has had a long-standing interest in international investment flows other than those it supports. In addition to various forms of cooperation with other multilateral development banks and with bilateral development agencies, the Bank has sought to promote private investment flows to strengthen the impact of its own resources and to support activities the World Bank may consider valuable but outside its own remit. To this end, it promoted the negotiation of the Convention Establishing the Multilateral Investment Guarantee Agency (MIGA).[13] Upon entry into force of the Convention in October 1985, MIGA became part of the World Bank Group.

The objectives of MIGA are 'to encourage the flow of investments *for productive purposes* among member countries, and in particular to developing member countries, thus supplementing the activities of' the World Bank, the International Finance Corporation and other international development finance institutions (italics added).[14] It is therefore not concerned with portfolio investment or the sale and purchase of other instruments derived from 'productive investment'. To achieve its objectives, MIGA issues guarantees against non-commercial risks in respect of investments in a member country that flow from other member countries. It also carries out 'appropriate complementary activities to promote the flow of investments to and among developing member countries'. Article 23 provides for a research function with respect to investment activities in developing countries and authorizes MIGA to 'promote and facilitate the conclusion of agreements, among its members, on the promotion and protection of investments'.

MIGA has limited programmatic means, involving the insurance of investments or guarantees for insurance (reinsurance) of investment, and a technical assistance programme that assists developing member countries in attracting FDI. With a capital stock of SDR 1 billion, MIGA's capacities are circumscribed.[15] 'A review of MIGA's existing portfolio indicates that about 30 percent of MIGA's current contracts are not project related, i.e. they are for investments in the financial sector. The remaining 60 percent [*sic!*] of MIGA's contracts are for various types of investors in projects. In MIGA's case, the "typical client" may be best characterized as a financial institution or a minority owner in a project' (italics deleted).[16]

In July 1990, the Development Committee of the International Monetary Fund and the World Bank published *Guidelines on the Treatment of Foreign Direct Investment*, specifying that 'these guidelines are not ultimate standards but an important step in the evolution of generally acceptable international standards which complement, but do not substitute for, bilateral investment treaties'.[17] The *Guidelines* articulate what may be deemed an interim consensus, albeit not a negotiated one, on a number of critical issues that will inform any broadly based multilateral investment regime. On the issue of 'performance requirements,' in effect obligations of investors, the *Guidelines* state: 'States will note that experience suggests that certain performance requirements introduced as conditions of admission (of investments) are often counterproductive and that open admission, possibly subject to a restricted list of investments (which are either prohibited or require screening and licensing), is a more effective approach.' It clearly recognizes the need for exceptions based on certain 'sectors reserved by law of the State to its nationals on account of the State's economic development objectives or the strict exigencies of its national interest'. Restrictions applicable to national investment on account of public policy, public health and the protection of the environment will also apply to foreign investment.

The fundamental issue of MFN and national treatment is addressed with caution, reflecting the range of practice to be observed in bilateral investment agreements. The key obligation is for each state to 'extend to investments established in its territory by nationals of any other State fair and equitable treatment according to the standards recommended by these Guidelines'. Rather than further developing the principles to be applied, the *Guidelines* proceed to cover a number of key issues, such as issuing of licences, transfer of funds and access of personnel. An entire section deals with expropriation and unilateral alterations or termination of contracts, with the focus primarily on the issue of compensation.

The Organisation for Economic Co-operation and Development (OECD) has a long-standing interest in FDI. This is readily understandable since the preponderance of such investment has occurred between member states of the OECD, and FDI in developing countries tends to have originated from OECD countries.[18]

The OECD is not normally a negotiating forum. Its principal activities concern the compilation and exchange of information between its member states. The OECD is the source of much information concerning FDI, some of which is further incorporated into the relevant UN documentation, and it has published a number of analytical studies on the issue. Over the years, the OECD has adopted a declaration and a number of decisions concerning international investment and multinational enterprises.[19] A 1976 *Declaration on International Investment and Multinational Enterprises* ties together all other OECD activities related to investment.[20] Directly linked to the *Declaration* are *Guidelines for Multinational Enterprises*.[21]

Following conclusion of the Uruguay Round, the Ministerial Council decided in 1995 to launch negotiations on a Multilateral Agreement on Investment (MAI). Given the difficulties that had been encountered in addressing the issue of investment in the GATT/WTO, there was clearly the intention to develop an instrument that would become the basis of a broader, global investment regime. In recent years, the OECD has extended its reviews of foreign direct investment to countries outside the organization, for example Ukraine, Chile, Argentina, and Brazil.[22] A number of these countries sat in on the MAI negotiations. In addition, a series of meetings was organized in parallel to the MAI negotiations to consult with the governments of the 'dynamic economies of Asia and Latin America'.[23] Given the existence of numerous bilateral and regional agreements and the broad consensus on the underlying principles, at least among the participants in the negotiations, framing the MAI was seen by many as primarily a technical task that could be completed fairly expeditiously.

The negotiations were conducted by a group of senior civil servants and were largely isolated from the regular business of the OECD, and from public debate. In March 1998, following an unprecedented international campaign triggered by environmental interests but supported by a wide range of groups that are sceptical about the processes of globalization and the distribution of its benefits, the MAI negotiations were put on a slower track, with no deadline. In October 1998, the process was abandoned entirely after France withdrew, mainly because it could not shield its cultural industries from the MAI rules. The newly installed German government also decided to press for 'social and ecological compatibility', which could not have been accommodated in the technical draft under consideration. The result is a lengthy, much-bracketed text that did not do much more than articulate the principles of MFN and national treatment, prescribe performance requirements, and provide for dispute resolution. In this regard, the text mirrors quite closely the typical bilateral agreements on investment concluded by many OECD countries. It also reflects many of the investment provisions of NAFTA.

The MAI is institutionally stunted. Trying not to pre-empt a later transfer to the WTO, or the creation of a separate entity for investment, the negotiators produced an institutional structure not unlike that of the original GATT, without a strong organizational base and with the narrowest of institutional resources. While the MAI text is lengthy and appears complex, it is in fact a relatively straightforward document, certainly much less complex than most environmental agreements, and the *Framework Convention on Climate Change* (*FCCC*) in particular.

In retrospect, launching the MAI negotiations as a technical process in the OECD without a clear political mandate can be seen to have been an error. By creating a negotiating process that was not integrated with the organizational structure of the OECD, the negotiators failed to perceive early warning signs. For example, an attempt by the OECD Environment Committee to engage the negotiators in debate was waved off. No thought was given to the problems that are likely to arise when bilateral agreements, which leave the parties in full control of the process, are turned into multilateral agreements, which create new international law and initiate a dynamic that must ultimately lead to new organizational structures.

The assumption that the principles underlying the GATT/WTO system are also appropriate for an investment regime has never been questioned. Even more seriously, the lessons from 50 years of struggling with the institutionally inadequate GATT appear not to have been learned. While the initial resistance to the MAI originated in environmental circles, it ultimately encountered a roadblock because no major party endorsed it enthusiastically. Once countries began to focus on its implications, the number of reservations grew so large as to nullify the effectiveness of the agreement. In other words, the negotiations collapsed because the original approach was flawed.

Investor obligations An entirely different approach to an international investment regime focuses on the obligations of investors. This reflects the fact that foreign investors acquire rights in the host country, and with these rights presumably also assume obligations.

The UNCTC *Code of Conduct on Transnational Corporations* never advanced beyond the draft stage. It is a defensive document. Its purpose has been defined as 'to maximise the contributions of trans-national corporations to economic development and growth and to minimise the negative effects of the activities of these corporations'.[24] The Code attempts to strengthen the position of (weak) developing countries in their relationship with large multinational corporations by defining a range of obligations, which would effectively transfer control over the investments to the receiving state. No investor is likely to undertake investment under such a regime.

The UNCTC *Code*, however, addresses a number of important issues, including the long-term nature of investments, the obligations of investors, the need for respect of national laws, and several issues relating to environmental management. Nevertheless, it has addressed these issues in a fashion that has served to further polarize the relationship between investors and the receiving countries. It does not adequately reflect the obligations already accepted by many developing countries in bilateral investment agreements. Moreover, its focus on 'multinational corporations' no longer corresponds to the complex reality of foreign direct investment.[25] It cannot be viewed as the basis for a multilateral regime, or even as setting the agenda for negotiations leading to such a regime.

The OECD has adopted *Guidelines for Multinational Enterprises*.[26] The *Guidelines* are a unilateral declaration on the part of governments, although they have been reviewed with representatives of industry prior to adoption. Adopted first in 1976, the *Guidelines* have been amended several times. The most significant change was the 1991 addition of a paragraph on the environment.

The *Guidelines* must be seen in relation to efforts within the United Nations to develop a *Code of Conduct on Transnational Organizations*. The UN launched this effort in 1974 with the establishment of the Commission on Transnational Corporations.[27] The OECD *Declaration* and *Guidelines* can be seen as an attempt to establish the boundaries for the UN effort. A comparison of the two instruments shows how differently they address a single agenda.[28] In recent years, the OECD approach is the only one to have survived. While the UNCTC approach was clearly incapable of gathering sufficient support, the issues it identifies still deserve serious consideration in any investment agreement that reaches beyond the OECD.

Dispute resolution Dispute resolution is an essential element of any international investment regime. For reasons outlined below, it will need to reflect the dynamic nature of an investment regime. There are, however, some precursors already in place.

In 1965, prior to the creation of MIGA, the World Bank had already established the International Centre for Settlement of Investment Disputes. The relevant Convention entered into force in October 1966 and has attracted 139 signatures.[29] At the request of parties to the Convention, the Centre establishes Conciliation Commissions or Arbitration Tribunals, drawing upon two panels to which each member country may name four persons, with the chairman of the Centre's Administrative Council adding a further ten. Commissions and Panels consist of one or more members, provided the number is uneven. The Tribunal shall decide a dispute 'in accordance with such rules of law as may be agreed by the parties'.[30] In other words, the Tribunals do not apply multilateral rules of law defined by the Centre, nor do they create a body of precedents that can be binding upon the parties, except insofar as they accept them at the time of the dispute.

ICSID has created the 'ICSID Additional Facility', which is available to countries not party to the ICSID Convention. The effect has been to universalize the coverage of the Centre, which is located at the World

Bank headquarters. The vice-president and legal counsel of the World Bank is the secretary-general of ICSID, so that ICSID is in practice a multilateral facility of the World Bank without the membership limitations of the Bank itself.

Over the years, ICSID has become widely accepted. Initially the countries of Latin America were sceptical about arbitration in investment disputes in general and the Centre in particular.[31] Recent bilateral agreements in Latin America have, however, also incorporated ICSID as a settlement mechanism, as have the Mercosur investment protocols.

It is generally assumed that the existence of ICSID, its inclusion in numerous bilateral and regional investment agreements, and the fact that its arbitration awards are not subject to judicial review have provided a powerful incentive for dispute avoidance. The fact that the Convention contains no rules of substantive law and no rules of conduct has presumably contributed to the willingness of parties to accept its jurisdiction.

Other investment agreements   The Framework Convention on Climate Change (FCCC) is in fact a multilateral regime on structural economic change and investment. It is, however, essential to keep in mind that the FCCC is establishing a complex set of rules governing international investments that reduce greenhouse gas emissions. These rules can be viewed as the nucleus of a specialized international investment regime, organized according to principles that are very different from those that govern trade regimes.

The FCCC is currently in a state of development. The Clean Development Mechanism, introduced through the Kyoto Protocol, establishes a new set of rules that are applicable to international investments. It is premature to speculate on the full extent of the FCCC's development. Should indications of serious climate change on a global scale prove well founded, the FCCC may yet become a comprehensive regime for the screening of investments to ensure that their negative impact on global climate change remains as limited as possible.[32]

A number of regional agreements also contain investment provisions. A 1996 compendium lists 31 regional instruments dealing with investment and eight free trade and regional economic integration instruments that contain investment-related provisions.[33] The variety of instruments is large, ranging from the treaties establishing the European Union to OECD instruments, and from the Community Investment Code of the Economic Community of the Great Lakes Countries (1982) to the APEC Non-Binding Investment Principles. Their common characteristic is that they rely on an existing organization to provide the means of implementation, rather than attempting to create an institutional framework of their own. Most important among the regional agreements is the North American Free Trade Agreement (NAFTA), because it represents a traditional trade agreement with investment provisions grafted on.

NAFTA has gone further than any other regional trade agreement in developing a set of specific rules for investment. It is based on the bilateral agreements developed by both Canada and the USA. Chapter 11 of NAFTA applies to foreign investors and investments (i.e. those from the other Parties) and its prohibitions on performance requirements and the injunction against encouraging investment by relaxing 'domestic health, safety or environmental measures' apply to all investments, both foreign and domestic.[34] The investment chapter articulates three principles, national treatment, MFN and a 'Minimum Standard of Treatment'. National treatment and MFN are required to be 'no less favourable than that it accords, in like circumstances, to investors of another Party'. The Minimum Standard is defined as 'treatment in accordance with international law, including fair and equitable treatment and full protection and security'.

Dispute settlement between parties under the chapter on investment falls under the general provisions of NAFTA governing the settlement of disputes between the parties. The chapter on investment contains a further set of rules governing the settlement of disputes between a party and an investor of another party.

A significant dispute between an investor and a party, which concerns environmental matters, has already been settled under NAFTA. Despite its salience for the NAFTA investment chapter – and for the development of an international investment regime – the details of the case of the Ethyl Corporation against the Canadian government remain largely obscure. The relevant documents have not been made publicly available, nor do they have to be. No account of the proceedings before the arbitration panel is available, and the settlement between the complainant and the Canadian government has not been published. Apart from articles in the specialist press,[35] a handful of Canadian newspaper articles,[36] and one item in *The Economist*,[37] reporting of the case has been limited. This limits the information base to press releases from the parties[38] and hostile comment from environmental and public health organizations,[39] hardly a satisfactory basis on which to develop international law.

MMT gas (methylcyclopentadienyl manganese tricarbonyl) was banned by an Act of the Canadian Parliament, adopted in early April. This special Act pre-empted, in the specific instance of MMT, a broad review of gasoline additives being undertaken under the Canadian

Environmental Protection Act (CEPA). Rather than issue an outright ban under CEPA, the Act prohibited the import and inter-provincial trade of MMT. Complicating matters, the (oil-producing) Province of Alberta opposed the law. Ethyl promptly filed a claim against the Canadian government for compensation under NAFTA, seeking damages for harm done to its subsidiary Ethyl Canada. While Ethyl Corporation has production facilities in Sarnia, Ontario, none of the public documents ever claimed that MMT was produced there or that it was the Canadian facilities that had been expropriated. Rather, the claim aimed at the company's reputation and business. Clearly, one of the major attractions of the NAFTA process was its speed. The case was settled within just over a year. In July 1998, the Canadian government rescinded the ban on importation and inter-provincial trade and agreed to pay Ethyl C$13 million in legal fees and compensation for lost profits. After initially professing confidence in its case, the Canadian government obviously concluded that there was substantial risk of losing it, without any appeal or further recourse. It is impossible to judge the accuracy of this assessment without access to the relevant proceedings of the arbitral tribunal, which was chaired by a German trade expert.

It remains difficult to assess the significance of the Ethyl case. The Canadian government settled, so as not to create a precedent – but in theory there is no 'case law' in the trade area, and the material success of the complaint establishes a sufficiently convincing precedent. Three salient points emerge:

- Under NAFTA, a private complainant can force a government to rescind a duly adopted legislative act.
- A NAFTA proceeding is faster and more conclusive than legal proceedings in national courts, which involve appeals and safeguards and consequently delay.
- If there was no production of MMT in Canada, the complainants appear to have used the investment provisions of NAFTA to oppose what can be viewed as a form of trade discrimination, effectively extending the notion of 'investment' to cover any investment anywhere and making the investor-state provisions of an investment agreement a back door to investor-state complaints on trade discrimination.

This dispute highlights one of the core problems in creating international investment regimes: by creating rights for foreign investors, the regime in practice contravenes the principle of 'national treatment' insofar as the foreign investors have rights the domestic investors do not have, namely to initiate proceedings against the host government.

Moreover, by creating a new body of international law – namely NAFTA itself – the new regime introduces a broad realm of uncertainty as to the proper interpretation of this law when inserted into the specific national context of a given country.

Unilateral action   A particular issue in relation to FDI is the continuance of 'home state' control, that is, the extent to which the country in which an investor is domiciled retains a measure of control over that investor's investments in another country. This is an issue where even investment insurance, which introduces the home state directly into the relationship between investor and host state, does not come into play. 'The home state continues to have an interest in the foreign investment after it leaves its shores, over and above its interest in the protection of the foreign investment through the principles of state responsibility and diplomatic intervention.'[40] This fact alone should warn negotiators that investments are different in character from trade in goods and will give rise to a range of issues beyond those encountered in multilateral trade regimes.

In the context of most regional and multilateral investment agreements, however, unilateral action represents a particular problem that needs to be kept within strict limits. Historically, European countries continued to exercise extensive unilateral control over the investments of their nationals in former colonies that had become newly independent. Indeed, protection of nationals and their investments has repeatedly been used as a reason for the use of force, mainly in the countries of Africa, by European countries.

The USA has sought to exercise control over investments of its nationals, and even over investments of nationals of other countries, in relation to its foreign policy. The examples are actually quite numerous and include embargoes against the People's Republic of China, measures seeking to undermine the construction of the Siberian gas pipeline to Western Europe, the Iran hostage crisis, and US efforts to impose an embargo on Cuba.[41]

As in the case of trade in goods, unilateral action is typically taken by a more powerful country against a less powerful one (although the USA has taken such action against a number of other major OECD countries). One purpose of a well-framed international agreement on investment must be to develop clear rules on the extent of home state interest in foreign investments, even when these are not subject to investment insurance schemes.

The Nature of an International Investment Regime

The social and legal context of investment The time frame of investments is widely divergent. Purely financial transactions – portfolio
investment in particular – can be extremely liquid, subject to purchase
and sale almost instantaneously. Within certain markets, there are now
specialist investors who purchase and sell investments within seconds.
The underlying social and legal relationship can only be described as
ephemeral.

Productive investments are, however, long term, measured in years
or even decades. The power plant that is built today may still be
operational a century later – well beyond the time when it is fully
depreciated – much modified but nevertheless in the same location and
often utilizing the same fuel. The forest that is cut down today may not
regenerate in two hundred years, and the farm or the plantation that is
likely to replace it will transform the landscape in which it is located. It
will be the object of changing crops and evolving practices from one
year to the next.

The underlying social and legal relationships established by longer-
term productive investments are significantly different from those involved in the sale of goods and the operations of portfolio investment.
Investment creates a complex system of rights and obligations, extending
into an indefinite future.

- Productive investments involve the purchase of contracts, which are
  open-ended, and of an indeterminate duration.
- An investor acquires a range of rights and obligations in the country
  where the investment is located. These may include rights to real
  estate, emissions rights, the right to contract with individuals and
  corporations, and the right to undertake financial transactions.
- An investor must accept obligations to respect the law of the juris-
  diction(s) in which the investment occurs and to contribute to the
  community in which the investment is located, for example by paying
  taxes.
- Frequently investments require infrastructure to ensure fair administration of the law, to provide for the needs of employees, and to
  ensure the availability of inputs and of transportation of output.

Viewed from a social and political perspective, trade in goods and the
making of a productive investment have hardly anything in common.
An international investment regime that serves the interests of all
parties concerned must reflect these differences at all levels, in the
principles it applies and in the details of its provisions.

An international investment regime that does not recognize the broader social dimensions of investment will contribute to the destruction of social and environmental values. It will defeat efforts to achieve greater sustainability. This is a claim that has frequently been made with respect to the entire process of 'globalization', and it is not always accurate. With regard to investment, however, the stakes are real. Constructing an international regime that is sensitive to a range of social, political and environmental variables linked to sustainability is a daunting task. It is impossible to encapsulate the demands of sustainability in a few operational principles, and there is still no firm international consensus with regard to many of the underlying issues.[42] An international investment regime must balance these considerations, and do so in a continuing manner that responds to changing priorities over time. To perform this balancing function, the investment regime must be dynamic in nature.

There is, however, a real possibility that an investment regime will be constructed largely as an extension of the WTO-based trade regime, with its emphasis on the removal of barriers to trade. The WTO regime is already in difficulty on account of its historic insensitivity to essential market disciplines, such as environment and labour standards, which it tends to view as 'barriers to trade'. Over the years it has become increasingly static, defending the essential trade disciplines rather than reaching out to address policy issues that may conflict with its own priorities. During the GATT era, there were institutional factors that dictated the static nature of the regime, since the GATT itself could not be amended or otherwise modified and the accretion of related agreements tended to render it inoperable. The advent of the WTO has resolved the problems of the associated agreements; it has not changed the static character of the regime. The entire GATT/WTO structure may collapse if an attempt is made to develop an investment regime that exhibits the institutional characteristics of the existing trade regime.

Four principles or norms are commonly identified as governing the 'extent and nature of liberalization' to be achieved in an international investment regime:[43]

- right of establishment;
- transparency;
- national treatment; and
- non-discrimination or MFN treatment.

This chapter will consider these four principles from the perspective of sustainable development, the only universally accepted criterion for evaluating the broader social and political significance of such a regime. It will argue that the principles of national treatment and MFN, the central principles of the GATT/WTO regime, will need to be balanced in light of two further principles:

- maintenance of competitive markets; and
- investor responsibility.

Moreover, it will maintain that the GATT/WTO dispute settlement system is inappropriate to the needs of an investment regime.

Some Issues for an Investment Regime that Promotes Sustainability

Investment is central to the prospects for achieving sustainability. Indeed, without new investment there is no prospect of replacing unsustainable infrastructure and production facilities. The sustainability concern in international trade is linked to the need for economic growth and the need to avoid the promotion of unsustainable practices. This goal can be achieved only if the incipient investment regime respects and incorporates the internationally agreed principles of sustainable development.

'Like' investments Most proposed multilateral investment agreements are based on MFN and national treatment, the central principles of the GATT/WTO system. The formulation of these principles revolves around the word 'like', that is, the obligation to treat 'like' products or investments equally. The difficulty arises from imposing an inflexible obligation in connection with an indeterminate term. The interpretation of the term 'like' represents in many ways the most serious conflict between the GATT/WTO regime and the requirements of sustainable development, about the need to distinguish between products produced sustainably and those produced unsustainably (for example, wood products from unsustainably harvested timber, or pond versus wild shrimp, or possibly agricultural products employing genetically modified organisms). Thus far, the GATT/WTO has steadfastly, and incorrectly, maintained that products cannot be distinguished by their mode of production.[44] GATT negotiators clearly recognize that a more determinate word (for example, 'identical' or 'same') would nullify the effectiveness of the central principles on which the multilateral trade regime rests. French and Spanish do not have a comparable word, so the translation renders 'like' as 'equivalent'. Equally, a less determinate word ('similar') would expose the regime to all kinds of arbitrary

discrimination. Unfortunately, interpreters of the GATT have, understandably, tended to emphasize the determinacy of the obligation over the need to interpret the indeterminate word 'like'.

If distinctions between 'like' products in international trade needed for ensuring greater sustainability of production have proved difficult to introduce, determining what are 'like' investments is sure to cause extraordinary difficulties. These difficulties are already reflected in the texts of most existing investment agreements, which refer not to 'like' investments but to MFN and national treatment 'in like circumstances'.[45]

An interpretative note proposed in March 1998 by the chairman of the MAI negotiations takes a first step towards discussing the difficulties surrounding MFN and national treatment for investments:

> National treatment and most favoured nation treatment are relative standards requiring a comparison between treatment of a foreign investor and investments and treatment of domestic or third country investors and investments. Governments may have legitimate policy reasons to accord differential treatment to *different types* of investments. Similarly, governments may have legitimate policy reasons to accord differential treatment as between domestic and foreign investors and their investments in certain circumstances, for example where needed to secure compliance with certain domestic laws that are not inconsistent with national treatment and most favoured nation treatment. The fact that a measure applied by a government has a different effect on an investment or investor of another Party would not in itself render the measure inconsistent with national treatment and most favoured nation treatment. The objective of 'in like circumstances' is to permit consideration of all relevant circumstances, including those relating to a foreign investor and its investments, in deciding to which domestic or third country investors they should appropriately be compared. (emphasis added)[46]

This text goes some way to identifying issues surrounding the concept of 'in like circumstances' in an investment agreement. It does not, however, address the issue of the interpretative process that is necessary to make these general observations effective. This process depends critically on the institutional capabilities of the international investment regime. An investment regime without an effective secretariat, with a dispute settlement process that relies on a changing group of arbitrators and without adequate public accountability, is liable to find itself embroiled in conflict within a very short time. The experience of NAFTA – which has modest institutional capabilities – is illustrative in this regard.[47]

The interpretative note does not, however, adequately address the temporal dimension of investment: circumstances can change over time, but a productive investment will continue and consequently 'like circumstances' can involve a great deal of variation.

National treatment   International investment agreements imply a trade-off between 'national treatment' and 'fair and equitable treatment' since the two tend to be used in a mutually exclusive fashion. This trade-off has never been clearly articulated. 'National treatment' implies a more affirmative right than 'fair and equitable treatment', even though both require a degree of interpretation – as acknowledged by the chairman's proposal cited above.

Environmental management is a dynamic activity, responding to growing knowledge concerning the environment and anthropogenic threats to it, as well as to changing perceptions concerning the seriousness of these threats. Moreover, environmental management is typically achieved through a 'package' of measures, involving standards, permits and licences on the one hand and economic incentives on the other. In addition, a complex structure of information and accountability, to management, to stockholders, to the authorities and to the public at large, represents a critical element of enforcement. Environmental management is always institutionally rich. The underlying reason for this complex approach is the difficulty in producing desired results in the natural environment, which responds to the laws of nature, through policy measures that affect only social behaviour. Consequently, the operation of environmental policy is always, and inevitably, indirect and subject to a degree of imprecision. To compensate for this imprecision, governments have been forced to use measures – such as command and control, incentives and informational obligations.

An added level of complexity derives from the continuous development of technologies designed to protect the environment. As these technologies become available, policy must adjust to reflect new capabilities.

Finally, the 'absorptive capacity' of the natural environment, such as it is, represents a scarce resource, to which there are no precisely delimited property rights, entailing a complex allocation process involving both public and private interests. Later 'like' facilities located in a watershed, or within the distribution range of atmospheric pollutants, must take into account the existence of prior emitters – which must in turn be subjected to new conditions to make room for new sources.

These characteristics of environmental management have two significant consequences: facilities are rarely 'like' from an environmental perspective; measures applied to otherwise 'like' facilities at different times are liable to be significantly different.

Environmental permits for installations as basic as coalfired power plants, one of the oldest forms of power generation, can differ widely from one facility to the next. Moreover, countries approach the problems arising from the dynamic and complex character of environmental management differently. A comparison of permits for coalfired power plants in the Netherlands and the Federal Republic of Germany showed that each facility had specific characteristics arising from the technology employed, the type of fuel, existing emissions, and shifting priorities of public policy, rendering comparisons virtually impossible. Moreover, administrative practice in the Netherlands allowed the continuous tightening of permits over many years; environmental management was in fact a continuous process of negotiation between the investor and the public authorities. In the Federal Republic of Germany, much more weight was placed on the long-term security of permits. New permits tended to be much more stringent – and much closer to the limits of current technologies – than in the Netherlands, but after several years, the requirements in the Netherlands tended to be more onerous than in Germany.[48]

There is no intrinsic reason why it should be impossible to determine what 'like' treatment is under these circumstances. It is, however, a complex process, which occurs continuously in all countries where equal treatment before the law is upheld. It is a demanding, continuous, dynamic process and raises questions about the institutional capabilities of an international regime. Certainly, an international agreement that provides for investor-state dispute proceedings needs to be developed with great caution. This is because it is liable to change in unpredictable ways the delicate balance that currently exists between investors and regulatory authorities within countries.

Faced with the challenge of developing appropriate environmental standards, issuing permits and licences and ensuring that all relevant measures have been complied with, environmental authorities in all countries are forced to engage in some form of selective enforcement. They must set priorities for enforcement action based on criteria such as the nature of the environmental threat, past history of a facility, or public pressure. Under these circumstances, determining what represents 'national treatment' can be a challenge.

One of the paradoxes of the principle of national treatment when applied to investments is that it does not put foreign and domestic investors on an equal footing – as it does when applied to goods in trade. It provides foreign investors with rights not enjoyed by their domestic counterparts, in addition to ensuring that they enjoy all the 'domestic' rights the latter have. In most countries, the grounds on which domestic actors can take a government to court are quite circumscribed. An international investment agreement – such as NAFTA – that gives private investors the right to initiate proceedings against host country governments establishes a new set of legal provisions for the benefit of foreign investors, which are not available to domestic investors.

Most favoured nation (MFN) treatment It would seem axiomatic that MFN is a strong principle for an international investment regime. Nevertheless, a number of issues concerning sustainability arise in this context as well. The potential conflict with the climate regime is fairly straightforward. In addition, the need for selective enforcement actions makes the environmental and management practices of the country of origin of an investor a matter for reasonable concern. Finally, with growing international integration of product chains, effective responsibility for certain environmental, and labour practices rests with the home country investor rather than with management in the host country.

The Framework Convention on Climate Change (FCCC) states: 'The Parties should protect the climate system for the benefit of present and future generations of humankind, on the basis of equity and in accordance with their common but differentiated responsibilities and respective capabilities. Accordingly, the developed country Parties should take the lead in combating climate change and the adverse effects thereof' (Art. 4). The notion of common but differentiated responsibilities and respective capabilities has caused a good deal of discussion. The existence in Annex 1 of a list of countries that have undertaken to limit their emissions of greenhouse gases, and the steady development of new instruments, such as activities implemented jointly (AIJ) and the clean development mechanism (CDM) launched by the Kyoto Protocol to the FCCC, introduce a range of new distinctions between countries, which are likely to result in distinctions between foreign investors from Annex I countries and all other investors. The result is a regime in which investments from developed countries in other Annex I countries would receive credits for reductions in greenhouse gas emissions while 'like' investments from non-Annex I countries would not. Similarly, investments from Annex I countries in non-Annex I countries would be favoured under the CDM, while investments between non-Annex I countries would not (see Chapter 6 for more details). The climate regime has not developed to the point where these effects are predictable. Nevertheless, an international investment regime needs to respect the requirements of the climate regime.

As governments confront foreign investors, in particular investors from 'offshore' investment countries, whose background is not or hardly known, it is not unreasonable to make certain distinctions based on the known requirements in the home market of the investor concerning environment and other factors of sustainability. An investment by a major enterprise from a country with rigorous environmental controls may attract levels of scrutiny different from those attracted by comparable investments from other countries, or investments from tax havens where there is no comparable activity at all. The last thing an international investment agreement should do is to be the investment equivalent of flags of convenience, which play such a central role in rendering international shipping, and its environmental performance and respect for labour standards in particular, almost impossible to control properly. This reflects the fact that, for example, in the provinces of coastal China, problems with FDI relating to the use of ozone-depleting substances are encountered primarily when the investors come from Hong Kong or Taiwan. It would certainly appear reasonable to seek particularly close control over their actions, whether or not this contravenes the principle of MFN. Similarly, it does not seem unreasonable to consider home country practices when awarding concessions to manage forests. For example, companies from Malaysia with a record of damaging forest practices have been acquiring forestry concessions in countries of Africa and Latin America. A government concerned about the future of its forests could reasonably be expected to impose additional conditions on the granting of such concessions under these circumstances.

Dispute settlement   The need for a dispute settlement process as part of any international investment regime is clear. Without such a process, the regime is unlikely to increase the calculability of risks to a significant degree. This dispute settlement process must, however, properly reflect the principles and the structures of the regime that is being created. Following the assumption that the investment regime would be based on the principles underlying the GATT/WTO regime, the obvious conclusion is that the dispute settlement process should be modelled on that of the WTO. This chapter has argued the inappropriateness of adapting GATT/WTO principles to an investment regime. It follows that the WTO dispute settlement process is also not suited for an international investment regime.

   Each factor that establishes the need for a dynamic investment regime – the long-term nature of investment and the consequent need to reflect change over time; the relationship between an investor and the investing country; and the difficulty of determining 'like circumstances' in an investment regime – also defines the characteristics of the dispute settlement process. It is not surprising that the issues that arise in developing a dispute settlement process for an investment regime are the same issues that have arisen in the confrontation between the (static) GATT/WTO system and the (dynamic) structures for environment and sustainability. The dispute settlement system for an investment regime must be open, accountable, capable of handling technical information and capable of balancing conflicting policy objectives. No such dispute settlement system yet exists at the international level, even though they exist at many other levels of governance.

Maintenance of competitive markets The continuing liberalization of investment is likely to create additional concerns relating to the maintenance of competitive markets, a vital factor for many developing countries dependent on the export of a limited number of commodities. The experience of the European Union has certainly been that as trade and investment are liberalized it becomes increasingly important to ensure that measures of competition policy are adopted at a level that corresponds to the dimensions of the relevant markets.

This is certainly not an immediate need, but it is a need that can be anticipated with some certainty. An international investment regime must be capable of addressing issues of competition policy at some point in the future. Presumably, addressing issues of competition policy is recognized for the significant institutional consequences that it entails; in practice, the institutional consequences for a regime in addressing investment are no less demanding.

Investor responsibility Investor responsibility has been the most important bone of contention in various attempts to address investment issues at an international level. The debate concerning investor responsibility was severely polarized in the 1970s and into the 1980s. The UN Code of Conduct dealt primarily with host state rights and investor responsibility; the OECD approach focused on investor rights, outlawed performance requirements and relegated what remained in the area of investor responsibility to a separate, less binding *Guide*. In the 1990s, the OECD approach appeared to prevail, as numerous countries moved towards liberalization of their capital markets. This trend was most pronounced in some countries of Latin America, Chile and Argentina in particular.[49] Mercosur adopted two protocols for Intra-Zone- and for Extra-Zone-originated FDI, whose provisions appear to be largely consistent with the OECD *Declaration* and subsequent OECD instruments.[50] As a result, attitudes towards FDI have shifted in many Latin American countries and have left the OECD approach as the only one currently under active consideration.

The failure of the UN Code of Conduct process should not obscure the fact that the polarization between the two approaches resulted in a situation where both tended to focus excessively on the particular issues being raised. In practice it is hard to see how an international investment agreement can be concluded that does not address issues of investor responsibility. A number of issues come readily to mind.

The most basic of all responsibilities is to respect the laws and regulations of the host country. Nevertheless, this simple statement requires some interpretation. Implementation of regulations is uneven in all countries, but more so in some than in others. Situations may arise where stringent regulations exist, for worker health and safety and environmental protection in particular, but where respect for these regulations is sporadic at best. Are foreign investors expected to comply with local practice, or are they to be measured by a more stringent standard?

In practice, the answer to this question may lie in the nature of the investment. Where the output from foreign investment feeds into a product chain in which the foreign investor plays a dominant role – for example, in electronics, in forestry or in mining – the foreign investor should be held to the most stringent interpretation of the law. Where the output of foreign investment is absorbed locally – for example, in some food processing or for transport services or power generation – the standards to be applied are essentially local in character, subject to the need of a foreign investor to protect its good reputation, or a global brand. Where obligations are international in nature – for example, relating to stratospheric ozone depletion, climate change or biodiversity – foreign investors must meet these international obligations. Of course the interpretation of obligations, which are variable in nature, requires a significant degree of institutional sophistication at the international level.

The issue of 'performance standards' has played an important role in the debate about international investment. These are obligations that may be linked to approval of investment and may differ, or seem to differ, from comparable requirements imposed on domestic investors. To the extent that certain foreign investors or their investments are different from domestic ones, there would seem to be no problem with imposing performance requirements on them that reflect their special status. In practice, however, using such differences to justify the imposition of standards is liable to give rise to serious conflicts within any investment regime, since it effectively transfers decision-making to the dispute settlement process.

The draft MAI takes a clear position on performance requirements. This list covers most of the issues commonly raised in connection with performance requirements. The implication of this provision is that governments will have to provide subsidies to investors if they want to impose performance requirements. It is also worth noting that governments may impose performance requirements on their nationals but would be prohibited from imposing them on foreign investors.

In practice, performance requirements are a reflection of the fact that the investor is acquiring significant rights in the jurisdiction of investment and that it is consequently normal for a negotiation to occur between the investor and the public authorities where the investment is to occur. This negotiation can cover a large range of topics, such as the provision of infrastructure, the right to utilize scarce natural resources such as air and water, the need to protect biodiversity and wildlife, employment, community development, and so on. The relationship between the parties can reflect a large number of different situations, ranging from dominance on the part of the investor to dominance on the part of the public authorities. It is essential that these negotiations be fair and equitable and that their outcome properly reflects the needs and interests of all parties concerned. While this may sound simple, it is not easy to achieve, but prior determinations of what topics are negotiable – which is the effect of the list in the MAI – do not resolve the underlying dilemma.

Transparency In dynamic regimes, transparency becomes a critical institution. Because of the processes of change involved in such regimes, and the continuing uncertainty about relevant facts and the identity of interested parties, dynamic regimes utilize transparency as an institution to ensure access to information and the participation of key actors, some of whom may not be known in advance. Closed regimes assume that all key parties are present or represented and that all material information will be available through them. That assumption is inoperable in a more dynamic situation and transparency is the appropriate response.

'Transparency' is not a process that gives rights to persons in an arbitrary manner. Apart from the basic right to know, and the obligation on members of the regime to be publicly accountable, actual participation can, and generally should, be made dependent on a showing of cause.

The instruments for transparency are by now well established, beginning with ensuring that all operational documents are made publicly available in a prompt manner, ensuring that meetings are public unless there are strong reasons for the respect of privacy, and providing avenues for persons outside the regime who may have an interest in outcomes to make their opinions heard for certain kinds of proceedings, dispute settlement in particular.

Conclusion: Institutional Capacity and an Alternative Approach

The controversy surrounding the draft MAI demonstrates the complexity of the issues that are at stake in an international investment regime. This chapter has argued that even apparently self-evident issues, such as MFN, national treatment or the elimination of performance requirements, will require much more careful consideration in a dynamic context. The reality of foreign direct investment has changed dramatically since the post-war era, since the elimination of fixed exchange rates, and since the debt crisis of the 1980s. More changes are to be expected, with the move towards more sustainable forms of production and the huge investments that implies, with the introduction of the euro, and with the crisis of international markets. Moreover, countries with ageing populations and capital surpluses will find that the only way to generate income for retirement-age people is through the investment of their capital surpluses in countries with younger populations.

These are issues quite unlike those addressed in the trade regime over the past 50 years. The process of reduction and elimination of tariffs, development of disciplines for non-tariff barriers and inclusion of further sectors such as services and agriculture in the GATT/WTO system never required a rethinking of the core principles of the regime. Only the agreement on intellectual property rights has created a new structure of rights and obligations – and it remains to be seen whether the resulting balance between monopoly for those who control intellectual property rights and the exposure of all other market participants to the commodification of their products is one that will prove endurable. It creates new sources of polarization and results in downward pressures on prices for goods taken from the natural environment, reducing the market valuation of the environment at a time when the move towards greater sustainability implies an increase in these valuations.

The GATT system was characterized by its institutional sparseness. The GATT itself could not be amended. Since the GATT secretariat was not an international organization, its functions were strictly limited. Implementation of the agreements under the GATT was multi-unilateral, that is it rested entirely with the countries involved, except in cases of conflict. Initially, this unusual structure served the trade regime well, since it was essentially a negotiating forum and breaches of the regime's disciplines could be tolerated as long as the general direction of trade liberalization was maintained. The dispute settlement process was central to the regime's success because it provided a slightly more dynamic source of interpretation than the few organizational structures.

It was not until the Uruguay Round that the institutional inadequacies of the GATT became so strong that a move was made to streamline the dispute settlement process and to draw together in a single organization, the WTO, the agreements that had proliferated with changing memberships under the GATT. The ensuing organization still has many of the characteristics of the GATT, in particular the unwieldy structure of councils and committees with unrestricted membership, the tendency to view all issues in terms of a negotiation, and the reliance on dispute settlement as principal means of implementation. Thus far, the WTO has proved no more adept than the GATT at addressing policy issues such as those relating to the environment, which require a balancing of conflicting goals. When all is said and done, the only priority that can be institutionally recognized is the need for trade liberalization.

An international investment regime deals with a much more complex agenda. It will need to be institutionally more sophisticated than the GATT/WTO system. One of the more surprising elements of the MAI text is the lack of attention to institutional needs. There appear to be a number of unarticulated assumptions concerning the appropriate institutional structure, derived from the GATT/WTO, which upon closer scrutiny turn out to be questionable. The MAI provides for a 'Parties Group', which resembles the GATT Council and the Contracting Parties of the GATT even more than the General Council of the WTO. It specifies that there will be a secretariat, which does not have any independent functions whatsoever.[51] This makes it a more virtual organization than even ICSID, which at least has a secretary-general who has certain specified, though modest, functions under the agreement.

The MAI is the first multilateral approach to investment that does not arise out of an existing institutional context, be it a regional trade agreement or the United Nations Centre on Transnational Corporations. It must consequently create the entire institutional structure needed to address the complex, dynamic agenda of investment, since it has no implied institutional background, like the European Union, NAFTA, Mercosur, the United Nations – or the OECD, for that matter. It reflects the assumption that the institutions that have served the trade regime will be adequate for an investment regime. Implicitly the negotiators must hope that the two will one day be merged. It is doubtful whether that will serve either the needs of trade liberalization or the demands of international investment in a satisfactory manner.

Reflecting the heavy reliance on the trade regime for its institutional inspiration, the draft MAI relies on dispute settlement as a means of implementation. The dispute resolution procedures of the MAI, modelled on those of the WTO and those of bilateral investment agreements, assume that the future task of the investment regime will be to apply an essentially immutable set of principles. The evolution of international investment over the past 50 years, and the necessary adjustment of public policy to go along with it, suggest, however, that the principal task of the investment regime will be the promotion of understanding the processes of investment and the maintenance of essential market disciplines – such as rules governing competition and the environment – that are necessary to ensure that investment serves overarching goals of public policy.

The only conclusion that can be drawn from an analysis of the MAI draft is that the negotiators assumed that investment is an act of nature and that the function of government is to stay out of its way. That is not an assumption that will promote more sustainable development.

All the attempts to address investment at the international level have also assumed that what is needed is a system of rules that must then be applied, much as in the trade regime, by governments and reinforced through a dispute settlement process. This is a fundamentally static view of the investment process and its function in economic and social policy. It makes no provision for the dynamic aspects of an investment regime, nor does it reflect the complex legal and contractual relationships between investor and host country that characterize foreign direct investment.

A better approach might draw some lessons from international environmental regimes that have faced the problem of addressing issues that evolve over time and consequently demand a dynamic international regime. The approach now well established in environmental regimes is to begin with a framework agreement that establishes basic institutions, creates an organizational structure and defines a continuing process designed to achieve certain articulated aims. Since negotiators cannot know which measures will ultimately be needed, most international environmental agreements are quite indeterminate as to the appropriate institutions that will be required. Over time, a body of evidence accumulates and additional measures can be adopted to ensure that the regime continues to move in the desired direction.

Conceivably, a framework agreement represents a better approach to addressing international investment policies. Such an agreement would outline a set of goals for the regime – including the goal of achieving sustainable development – as well as a process to explore necessary steps towards those goals. The result would be an incremental regime, capable of responding to emerging needs and adapting to changing practices in international investment.

The central challenge of any international investment regime will be to maintain a high level of predictability while retaining essential flexibility. In many ways, that resembles the dilemmas of the institutions that govern the monetary system. While they need to be highly stable and predictable over long periods of time, in moments of crisis they must act decisively even if this breaks with what appeared to be well-established precedent. The challenge facing an international investment regime is not as daunting as that confronted by monetary authorities. Its purpose is, after all, to ensure the highest possible level of calculability of private economic risks in the investment process while ensuring that the overriding goals of public policy – sustainability, human rights and the vitality of communities – are respected. That does not sound like a goal beyond the reach of international society as it has evolved over the past decades, but it is a goal that can be attained only if the investments regime reflects an adequate institutional base and the capability to balance conflicting goals of policy.

Notes

1. Adjunct professor, Dartmouth College.

2. Fitzgerald et al. 1998.

3. UNCTAD 1996a: iv. Fitzgerald et al. (1998: 28) speak of 1,600 bilateral treaties, without providing additional sources.

4. UNCTAD 1996a: 115–258.

5. See below.

6. See below.

7. The United Nations Commission on International Trade Law (UN-CITRAL) is an organ of the UN General Assembly. Established in 1966, UNCITRAL promotes harmonization and unification of commercial law. It developed the Convention on the Recognition and Enforcement of Foreign Arbitral Awards, which entered into force on 7 June 1959 (www.un.or.at/uncitral/english/texts/arbconc/58conv.htm – 3 January 1999). UNCITRAL has, by resolution of the General Assembly, established Arbitration Rules (www.un.or.at/uncitral/english/texts/arbconc/arbitrul.htm – 3 January 1999) and Conciliation Rules (www.un.or.at/uncitral/english/texts/arbconc/concirul.htm – 3 January 1999).

8. Sornarajah 1994: 187–223.

9. Sornarajah 1994: 223.

10. UNCTAD 1987: 369.

11. WTO 1995: 163–7.

12. Article 9 reads: 'Not later than five years after the date of entry into force of the WTO Agreement, the Council for Trade in Goods shall review the operation of this Agreement and, as appropriate, propose to the Ministerial Conference amendments to its text. In the course of this review, the Council for Trade in Goods shall consider whether the Agreement should be complemented with provisions on investment policy and competition policy.'

13. UNCTAD 1996b: 213–43.

14. Article 2.

15. Multilateral Investment Guarantee Agency, 'MIGA: The Mission and the Mandate'. www.miga.org/welcome (2 January 1999).

16. The Multilateral Investment Agency, 'Draft Environmental Assessment and Disclosure Policies and Environmental Review Procedures', www.miga. org/welcome.htm (2 January 1999: 2).

17. UNCTAD 1996a: 247–55.

18. Dunning 1997: 20–1.

19. OECD 1997.

20. Ibid.

21. Ibid.

22. Reviews of foreign investment are published by the OECD: Brazil (1998); Ukraine (1997); Chile (1997); Argentina (1997).

23. OECD 1995.

24. Cited in Sornarajah 1994: 195 as part of the Preamble. This text is, however, reproduced in 'Draft United Nations Code of Conduct on Transnational Corporations', in UNCTAD 1996b: 161–80, which reflects the status of negotiations as at 1986.

25. See above.

26. OECD 1997: 103–10.

27. ESC, Resolution 1913 (LVII), 1974.

28. CUTS 1998: Annex 2. This useful paper does not use the latest version of the OECD *Guidelines*.

29. World Bank, 'An overview of ICSID', www.worldbank.org/html/extdr/icsid/html (2 January 1999: 1).

30. Article 42 (1).

31. Broches 1985: 83, 77.

32. On the development of the Climate Convention, refer to Chapter 6 and also see www.unfccc.de, which contains all the relevant documents.

33. UNCTAD 1996a.

34. Art III.4.1: 'Nothing in this Chapter shall be construed to prevent a Party

from adopting, maintaining or enforcing any measure consistent with this Chapter that it considers appropriate to ensure that investment activity in its territory is undertaken in a manner sensitive to environmental concerns.' Art. III.4.2: 'The parties recognize that it is inappropriate to encourage investment by relaxing domestic health, safety or environmental measures. Accordingly, a Party should not waive or otherwise derogate from, or offer to waive or otherwise derogate from, such measures as an encouragement for the estab-lishment, acquisition, expansion or retention in its territory of an investment of an investor. If a Party considers that another Party has offered such an encouragement, it may request consultations with the other Party and the two Parties shall consult with a view to avoiding any such encouragement.'

35. Hilleman 1998, 13.

36. 'Additive fears lack clout', *Globe and Mail*, 20 July 1998. George Monbiot, 'Running on MMT', *Guardian* (London), 1998.

37. 'The sting in trade's tail', *The Economist*, Vol. 347, No. 8064, 18 April 1998: 70–1.

38. Ethyl Corporation, 'Bill C-29 creates serious problems for federal govern-ment', www.ethyl.com/news/2-7-97.html (3 January 1999); Ethyl Corporation, 'Ethyl files claim to seek not less than US$250 million', www.ethyl.com/news/4-17-97.html (3 January 1999); Ethyl Corporation, 'Ethyl welcomes government of Canada decision', www.ethyl.com/news/7-20-98.html (3 January 1999); Appleton & Associates, 'First-ever lawsuit against Canadian government using NAFTA investor–state process brought', press release dated 9 October 1996 (cited in Sforza and Valliantos (see note 39 below), unavailable on the Appleton & Associates website on 3 January 1999).

39. Michelle Sforza and Mark Valliantos, 'NAFTA & environmental laws: Ethyl Corp. v. Government of Canada' (April 1997), www.globalpolicy.org/socecon/environmt/ethyl.htm (3 January 1999); Friends of the Earth, 'Canada withdraws environmental law after NAFTA challenge', press release dated 20 June 1998; Canada Safety Council, 'Secret panel to decide on future of contro-versial additive', www.safety-council.org/mmt.htm (3 January 1999); Environ-mental Defense Fund, 'EDF dismisses Ethyl's claim of trade discrimination on MMT; cites new survey showing over 85% of US Oil Co.'s reject use of MMT gas additive', press release dated 10 September 1996, www.edf.org/pubs/News Releases/1996/Sep/ (3 January 1999).

40. Sornorajah, (1994: 145).

41. For a comprehensive analysis of US sanctions see Hufbauer and Schott 1985.

42. See IISD 1994.

43. Ostray 1997: 11.

44. For example, GATT 1994 stipulates: 'any advantage, favor, privilege or immunity granted by any contracting party to any product originating in or destined for any other country shall be accorded immediately and uncon-ditionally to the like product originating in or destined for the territories of all

other contracting parties' (Art. I.1) and 'The products of the territory of any contracting party imported into the territory of any contracting party shall not be subject, directly or indirectly, to internal taxes or other internal charges in excess of those applied, directly or indirectly, to like domestic products' (Art. III.2).

45. For example, the draft MAI uses the following language: 'Each Contracting Party shall accord to investors of another Contracting Party and to their investments, treatment no less favorable than the treatment it accords [in like circumstances] to its own investors and their investments with respect to the establishment, acquisition, expansion, operation, management, maintenance, use, enjoyment and sale or other disposition of investments' (Art III.1). 'Each Contracting Party shall accord to investors of another Contracting Party and to their investments, treatment no less favorable than the treatment it accords [in like circumstances] to investors of any other Contracting Party or of a non-Contracting Party, and to the investments of investors of any other Contracting Party or of a non-Contracting Party, with respect to the establishment, acquisition, expansion, operation, management, maintenance, use, enjoyment and sale or other disposition of investments' (Art III.2). The bracketed text indicates some uneasiness about the term 'in like circumstances', at least on the part of some negotiating parties.

46. 'Environment and Labour in the MAI. Chairman's Proposals: March 1998' (updated 15 May 1998).

47. See the discussion of the Ethyl case above.

48. Von Moltke et al. 1985.

49. Agosin 1995.

50. Colombi and Podrez 1997: 115–25.

51. 'The Parties Group shall be assisted by a Secretariat' (Art. XI.7).

Bibliography

Agosin, M. R. (1995), *Foreign Direct Investment in Latin America*, Washington, DC: Inter-American Development Bank.

Broches, A. (1985), 'The experience of the International Centre for Settlement of Investment Disputes', in J. Rubin and W. Nelson (eds), *International Investment Disputes: Avoidance and Settlement Studies in Transnational Legal Policy No. 20*, St. Paul, MN: West Publishing.

Colombi, P. and J. Podrez (1997), 'Mercosur protocols for foreign investment promotion and protection', in OECD, *Investment Policies in Latin America and Multilateral Rules on Investment*, Paris: OECD.

CUTS (Centre for International Trade, Economics and Environment) (1998), *Multilateralisation of Sovereignty: Proposals for Multilateral Frameworks for Investment*, Discussion Paper No. 79/80, Jaipur: CUTS.

DFID (1998), *The Development Implications of the Multilateral Agreement on Investment*, a report commissioned by the Department for International

Development, Finance and Trade Policy Research Centre, University of Oxford.

Dunning, J. H. (1997), 'The advent of alliance capitalism', in J. H. Dunning and A. Hamdani (eds), *The New Globalism and Developing Countries*, Tokyo: United Nations Press.

Fitzgerald, E. V. K. et al. (1998), *The Development Implications of the Multilateral Agreement on Investment*, a report commissioned by the Department for International Development (UK), Finance and Trade Policy Research Centre, University of Oxford.

Hilleman, B. (1998), 'Canada capitulates on MMT, settles with ethyl', *Chemical and Engineering News*, Vol. 76, No. 30.

Hufbauer, G. C. and J. Schott (1985), *Economic Sanctions Reconsidered: History and Current Policy*, Washington, DC: Institute for International Economics.

IISD (1994), *Principles for Trade and Sustainable Development*, Winnipeg: International Institute for Sustainable Development.

von Moltke, K. and F. Meiners (1985), *Rechtsvergleich deutsch-niederländischer Emissionsnormen zur Vermeidung von Luftverunreinigungen Teil 1: Bundesrepublik Deutschland; Teil 2: Niederlande; Teil 3: Tabellen. (Teil 1* also in Dutch: *Rechtsvergelijking van duits-nederlandse emissienormen ter bestrijding van luchtverontreiniging)*, Bonn: Institute for European Environmental Policy.

OECD (1995), *Countries and Dynamic Economies of Asia and Latin America*, Paris: OECD.

— (1997), *The OECD Declaration and Decisions on International Investment and Multinational Enterprises: Basic Texts*, OECD Working Papers, Vol. 5, No. 7.

Ostray, S. (1997), *A New Regime for Foreign Direct Investment*, Washington, DC: Group of Thirty.

Sornarajah, M. (1994), *The International Law on Foreign Investment*, Cambridge: Cambridge University Press.

UNCTAD (1987), *Ministerial Declaration on the Uruguay Round: Papers on Selected Issues*, New York: United Nations.

— (1996a), *International Investment Instruments: A Compendium. Vol. I: Multilateral Instruments*, New York: United Nations.

UNCTAD (1996b), *International Investment Instruments: Compendium. Vol. III: Regional Integration, Bilateral and Nongovernmental Instruments*, New York: United Nations.

WTO (1995), *The Results of the Uruguay Round of Multilateral Trade Negotiations: The Legal Texts*, Geneva: WTO.

. . . .

PART IV

*Proposals for a Policy Response
by the South*

. . . .

CHAPTER 8

An Agenda for Change

Mark Halle[1]

§ The following have been well established in this volume: first, that the links between trade and environment are both strong and significant; second, that the links between trade and environment are complex. It has been shown in this volume and elsewhere that the trade and environment interface includes both threats and opportunities; that trade liberalization can be good for the environment, just as it can be bad, the difference lying in the extent to which the policies governing trade and environment are harmonized and coherent. The same clearly applies to the effects of environmental measures on trade.

The benefits to the South will come from getting the balance right, not from ignoring the issue. It would be dangerous to stay out of the game if the game has to be played, and increasingly it seems as though it will have to be played. In view of this, the South will have to become much more active.[2] Even though Southern interests are different from those of the North, the South has environmental interests that need to be defended.

The WTO High-Level Symposium on Trade and Environment (Geneva, 15–16 March 1999) leaves no doubt that environment is becoming important for the WTO. It is a clear political priority for the North due to the steadily growing public demand for environmental quality and performance. As such, it can be regarded as a negotiating asset for the South, but only if it is in the WTO, as part of the package.

The negotiations aimed at agreeing to a Bio-safety Protocol under the Convention on Biological Diversity in Cartagena (Colombia) demonstrated the beginnings of a new politics of the developing countries, where they cluster around interests and not along North–South lines. The policy of anger and resentment could be giving way to a politics of strategic defence of central interests.

This means that countries in the South will have to articulate their legitimate environmental interests and gain a clear understanding of how these are undermined by the trading system. At the same time, they must understand how a judicious use of the trade rules can advance these interests. This will mean developing the capacity not only to track and analyse these issues, but also to defend national interests effectively in those fora where major decisions are taken.

In observing country X in the South, we can say we are currently at point A. Let's imagine a point B, where country X has a fully developed capacity to follow and participate in the debate on trade and environment, to analyse and articulate its legitimate national interests in the trade and environment interface, and to advance these interests successfully in international fora where trade and environment are debated and major decisions taken. How can we best construct a road map that will move country X rapidly, effectively and affordably along the path from A to B? I will make suggestions under seven headings: information, networking, research, capacity development, institutional arrangements, legitimacy and regional cooperation. Most focus on domestic preparation, because if that is not in place, the rest won't work. Some of the suggestions set out below may strike one as simplistic, but progressing at a steady pace by taking practical and affordable steps is the key.

Agenda

Information

RESOURCE CENTRES Establish one or several basic collections of documents, books and articles relating to the various aspects of trade, environment and sustainable development.[3] These collections should be kept current, and should be openly available to researchers, the media and civil society. Special attention should be paid to resources contained in the recommended readings service of the International Centre on Trade and Sustainable Development (ICTSD).

NET SEARCHES The capability should be established in one (or several) institutions to track the various internet sites relating to trade, including the IISD's Linkages and others, and to ensure that all interested members of a trade and environment network (see below) have access not only to the information, but also to relevant ListServes and other means of participating in the internet-based debates, networking and advocacy. The ICTSD weekly *Trade News Bulletin* and its monthly *Bridges* newsletter are of particular relevance.

NEWSLETTER A trade and environment newsletter should be produced on a regular basis – monthly at least, if not weekly or bi-weekly. This newsletter should draw on a wide range of internet and documentary sources, using as a grid those issues and events most of interest and significance to the country. Its aim should be to select the best and most relevant information and make it easily accessible to a wide public – thus providing a valuable short-cut to information. Progressively, it should have an increasing component of original content.

SEMINARS AND WORKSHOPS NGOs and policy institutions should multiply the number and variety of informal occasions for information-sharing, awareness-raising and debates. They should seek opportunities for 'brown bags', seminars, workshops, occasional lectures and press briefings.

Networking The patient work of constructing a trade and environment network should be undertaken. This means initiating a database of names and contact numbers. The principal tool for networking will, at least initially, be the newsletter and other information services, so that the readership of the newsletter will gradually expand and be refined until it encompasses all those interested in trade and environment issues. Civil society organizations should also develop a network, or a component of the network, having as one of its principal targets the development and strengthening of NGO advocacy capability.

Research It is essential that the countries strengthen their indigenous research capacity to identify the key environment issues in the trade context, to pinpoint and refine legitimate national interests, and to detect the levers that will bring about change most effectively. The presentations of the cotton and leather industry case studies in this volume provide examples of how this approach can function. The above process should lead to the development of an agreed national research agenda on trade and sustainable development.

Research should in particular focus on the *positive agenda* – for example, the identification of new niche markets for environmentally sound products, or the potential represented by the trend towards *green procurement* in many countries, coupled with the trend towards liberalization of government procurement. Another important area of research concerns the impact of WTO regulation on national environmental legislation.

As has been documented in various chapters in this volume, many countries in the South suffer from *bad faith implementation* of the current WTO and other multilateral agreements (for example, the end-loading of the compliance period for dismantling textiles quotas). This phenom-enon can be attacked, but only on the basis of specific examples and sound research. The same is true of the problems caused by the fact that agreed international norms and standards – such as those on food safety, health and environment – are largely set by Northern countries on the basis of Northern realities. Once again, this situation will change not on the basis of general unfairness, but on the basis of scientifically demonstrated issues and scientifically sound proposals.

The purpose of research on trade and environment must not simply be to push back the frontiers of knowledge, but must be centrally geared to the development of clear policy options. The research agenda should not simply focus on national issues, or national impacts of international issues. Southern countries should participate actively in the research networks developing around trade issues regionally and internationally. The IISD-based Trade Knowledge Network provides an excellent example. It is constructing a network of policy institutions, both North and South, and is assisting with information and internet access and the identification of common issues. As a global network with both developed and developing country participants, it is uniquely able to bypass the trading system and find policy solutions that actively contribute to the achievement of sustainable development through the trading system.

Capacity development Beyond general awareness-raising, workshops and short courses should be organized, based for example on key issues coming up on the WTO agenda, on individual WTO agreements (such as TRIPs, TRIMs, SPS, TBT, GATS and the Agreement on Subsidies and Countervailing Duties), or on key export commodities (such as cotton and textile, leather and hides, fisheries or plant varieties). As countries begin to identify the key trade and environment issues in the national interest, these should of course provide the basis for capacity-building events.

Awareness-raising and capacity-building events targeted at the cor-porate sector are also necessary. The cooperation of the World Business Council for Sustainable Development might be sought in this regard. Creative thought should be given to other specific events that can easily be organized, such as, for example, a regular lecture series.

Institutional arrangements As has been mentioned in various chapters of this volume, strengthening openness, debate and participation in the formulation of national trade policy and national positions on trade issues will in turn contribute to strengthening coherence among different government policies, thereby making them more effective. A national forum on trade and environment should be created, where government, business and civil society can regularly discuss trade policy issues relating to the environment. Debates in this forum should aim to strengthen and confer legitimacy on government policies and the positions taken by government in international negotiations. The example of India is notable in this regard.

A national forum should not preclude the establishment of more focused or specialized sub-fora on specific issues on the trade and environment agenda. An inter-agency task force on trade and environment should be created, with the very specific task of reviewing the different policies relating to trade and environment for harmony and coherence.

The above-mentioned network on trade and environment should organize and operate a briefing facility aimed at providing regular briefs to government departments and delegates on forthcoming issues and of contributing to the preparation of national positions.

Inspired by the examples of countries such as Singapore or India, countries should deliberately seek to develop a cadre of negotiators who follow the trade and environment debate and the negotiations on a steady and consistent basis, thus building up experience, understanding and sophistication, and thereby greatly enhancing the effectiveness of their voice on these issues internationally.

Governments should name a trade and environment focal point – with a watching brief on potential trade-related environmental requirements being developed internationally – for instance, new labelling or certification schemes, new norms and standards and new government procurement criteria. Finally, the focal point should see to the incorporation of trade and environment considerations in relevant national plans, strategies and work programmes.

Legitimacy The policy on trade and environment must genuinely represent the national interest, and not simply the interests of an elite of export-oriented industries. This will require the process of policy development to be made more transparent, and involve appropriate participation and input from relevant stakeholders. The national forum mentioned above could serve as the vehicle for enhancing the legitimacy of national trade policy. Legitimacy requires not only transparency, but also accountability. Mechanisms must be found for public reporting on, and justification of, positions taken by national representatives in international fora.

Regional cooperation On their own, many Southern countries would have difficulty in advancing their views when faced with powerful trading blocs such as the European Union or powerful economies such as that of the USA. In addition to the interest-based clusters mentioned above, there is clearly scope for greater regional cooperation. Such efforts need to be mirrored by regional cooperation among NGOs, policy institutes, corporate groups, and so on.

Conclusion

As has been stressed repeatedly in this volume, there is nothing to be gained and a great deal to be lost by adopting a policy based on anger and resentment. The process of multilateral rule-making on trade and multilateral environmental cooperation is moving ahead. If Southern countries do not participate actively, they will find themselves increasingly marginalized.

Southern countries must get into the game, understand its rules, and make them work for them, even if they are currently unbalanced and unfair. A Tanzanian minister once wryly remarked: 'If I must get into the ring with Mike Tyson, it is not much of a consolation that the rules are the same for both of us.' But we do have to get into the ring, so it seems sensible to begin with some effective coaching and training, a better diet and a body-building programme. It might eventually get us through the next round. The next chapter of this volume addresses this issue in more detail.

Notes

1. European director, IISD.

2. See also suggestions in this regard in the concluding chapter of this volume.

3. My concern is much more with the broader trade and sustainable development interface rather than the narrower trade and environment interface addressed by this volume. See also Najam's support for this position in Chapter 9.

· ·

Towards a New Global Agenda on Trade and Sustainable Development

Adil Najam[1]

§ Earlier chapters in this book amply demonstrate that the links between trade issues and environmental concerns are deep, but complex. Moreover, this volume adds to a now robust literature that convincingly argues that developing countries have legitimate and significant apprehensions about the general direction of global debates on trade and environment (see Nath 1997; Shahin 1998; South Centre 1998; Raghavan 1999, Youssef 1999). In particular, the developing countries of the South fear that the imposition of Northern environmental concerns on the international trade agenda will a) distract from other more pressing Southern concerns and b) open the floodgates for so-called 'green protectionism', which will be particularly detrimental to developing country products and services.

At the same time, however, the writing on the wall is becoming increasingly clear to the developing countries. There is a very clear sense in the policy pronouncements from the North as well as in the dominant academic discourse that even if it is possible to delay the induction of a stronger environmental focus within the global trade regime, it cannot be postponed indefinitely (see Whalley 1996; Jha et al. 1997; Brack 1998; Uimonen 1999; Sampson 1999). As von Moltke (1999: 7–8) points out, 'the current political situation in Europe and North America reflects a reality that will prove inescapable' and 'it is hard to imagine any future trade agreement ... which does not seriously address environmental concerns'. If anything, the events at the Third Ministerial Meeting of the World Trade Organization (WTO) suggest that the issue will have to be dealt with sooner rather than later. The relevant question for the South, therefore, is no longer *whether* environmental concerns should influence trade rules, or even *when*, but *how*. This chapter will focus on exactly this last question: *How should the* developing countries of the South respond to trade and environment issues in future international negotiations?

The chapter addresses the issue from a decidedly Southern perspective. Its point of departure is a twin premise that follows directly from the evidence presented in the preceding chapters and from the larger literature. The first part of the premise is rooted in a strong belief that the general thrust of Southern concerns on trade and environment is valid and justified, both politically and substantively.[2] The second part emanates from an acute realization that the South can no longer afford to ignore the trade and environment link. If the developing countries disengage themselves from this discussion, they will do so at their own peril.

From this point of departure, the chapter sets out to propose a proactive and positive strategy for the South for future international negotiations – one that seeks the achievement of sustainable development in the South through active engagement of the developing countries in the trade and environment discourse. In doing so, however, it is both strategically and conceptually important to place the trade and environment question within the context of other issues that now frame international trade discussions and may be of equal or greater importance to the South. The chapter begins, therefore, with a quick survey of the spectrum of trade-related issues that are of concern to the collective South. It then reviews options for advancing the WTO negotiations stalled at Seattle in a manner that could meet the South's interests. The chapter then goes on to the question of why the South should adopt a proactive negotiation agenda on trade and environment and, finally, what the key elements of such an agenda might be.

International Trade Regime: Key Concerns for the Collective South

The debate on trade and environment does not happen in a vacuum. Throughout the 1990s, a number of issues took prominence in the international trade regime of which this was just one. In particular, the broad mandate set out in the Marrakesh Declaration at the end of the Uruguay Round, the formation of the WTO, the process of rapid globalization, and the continued expansion of the trade agenda have had an overwhelming effect on the developing countries (South Centre 1998; Raghavan 1999). Bombarded by a plethora of new issues – and endemically constrained by a comparative disadvantage in terms of a major expertise and resource deficit – developing countries are justifiably worried that the focus on trade and environment will necessarily dilute the attention afforded to other issues on the agenda.

Many developing country analysts, such as Raghavan (1999), consider collective bargaining by the South to be a logical antidote for this situation. Similarly, in its important report on the subject, the South Centre (1998: 19) suggests that

> developing countries need to play a more incisive role in the WTO in pursuit of their interests. It is essential that they engage in discussions and consultations among themselves at different stages of their WTO work in order to evolve common perceptions and, on this basis, formulate specific proposals and positions to advance their objectives. Such efforts will require thorough technical preparation in each area of work.

Others, particularly in the North, consider this call for Southern solidarity on trade issues, and on trade and environment issues in particular, to be a cause for concern. Whalley (1996: 92), for example, argues that:

> The trade and environment issue raises the probability that developing countries will resume a collective approach to trade issues, and that a North–South divide could re-emerge on these issues. The potential for such a conflict is large. The cohesiveness of the interest in the developing world is established. The desire for compensation rather than retaliation is very strong. Moreover, the sense of being at a disadvantage to developed countries would be aggravated by the extensive use of measures to restrict trade on environmental grounds by developed countries.

From a negotiation theory point of view it is, of course, possible, feasible and desirable for countries within and across the North–South divide to align with other countries having similar interests. Historically, such alliances have not been uncommon in the trade arena. In fact, developing countries have tended to be less eager to negotiate as a group in the GATT/WTO context than they have in UN-based forums such as UNCTAD or in various multilateral environmental agreements. However, the political fact of the matter is that the North–South dimension is becoming increasingly relevant to international trade discussions. In some measure, this is because developing as well as industrialized countries have tended to view trade and environment in particular as very much a North–South issue. While no one would wish for needless polarization in international negotiation, greater Southern solidarity could, in fact, make international negotiation more effective for at least two reasons. First, it serves to convert a large-N negotiation into a small-n negotiation, since the process of alliance and coalition building effectively reduces the number of parties (or, at least, positions) in the dialogue. Second, it allows developing countries that individually lack negotiation resources and expertise to pool their resources and expertise together. In essence, more coordinated negotiation by the developing countries would not only reduce the number of positions around the negotiation table, but ensure that the positions are better thought out and prepared.

Indeed, in calling for improved coordination between developing countries this chapter does *not* seek North–South polarization or 'bloc politics'. In fact, it envisages a number of issues where alliances between Southern and Northern countries and groups could, and should, blossom on the basis of mutual interests. On other issues, however, where a strong commonality of interests exists between developing countries – particularly issues related to development, including just and sustainable development – it is to the benefit of the developing countries, and of the global trade regime, that the South adopts a more coherent, cohesive and coordinated negotiation strategy.[3]

The rest of this section will quickly summarize some of the signature issues and interests that are of concern to developing countries as a group, and identify the key arguments around which a coordinated Southern negotiation strategy might be constructed.[4]

Trade liberalization and growth The South is being given the impression that there is overwhelming evidence that trade liberalization necessarily generates growth. This, in fact, is not true. The cross-country models used to establish these results are flawed because a) they assume a high degree of similarity across countries, which is clearly not the case; b) the results tend to be very sensitive to country sample, time period and specificities of the model (see Bilginsoy and Khan 1994); and c) in focusing on protectionism, studies have tended to underestimate the importance of internal competition policy, which is much more relevant.[5] Developing countries, therefore, should not be defensive when confronted with the grandiose theology of liberalization.

Competition policy For the South this should mean: a) oligopolistic practices of large multinational corporations (MNCs); and b) anti-dumping, subsidies, quotas and other non-tariff barriers that act as a restraint to trade of Southern exports. The North may want to focus on 'market access' for their MNCs, but the South should argue that since they do not have MNCs that can be granted reciprocal benefits, this becomes a very one-sided view of competition policy.

Dumping This is a key issue for the South since the North's use of the anti-dumping mechanism is often suspect. The South needs to build its analytical capacity to fight such cases. In particular, it should be able easily to show that its exporters, who are subject to quotas on commodities such as yarn, have little incentive to dump since they will not enhance their market share by doing so.

Agriculture This is, of course, a principal substantive concern for many developing countries. The South has sought to make it quite clear that currently agricultural export is highly monopolistic and dominated by Northern MNCs. Thus, there is no 'free trade' for many agricultural commodities, and creating free trade in this sector should be a priority. Developing countries should also argue for self-sufficiency in basic foods to preserve food security.

Sanitary and phyto-sanitary measures and technical barriers to trade Among the many concerns of common interest in this area, three stand out in particular: a) since many developing countries cannot be represented at standard setting meetings where decisions-by-vote are taken, a system of proxy votes should be allowed where votes can be transferred to other like-minded countries; b) due to the high cost of certification standards such as the ISO 14000 series, Southern countries should be provided technical assistance to develop their own internationally recognized bodies for certification; and c) clean industrial technologies should be promoted in accordance with *Agenda 21* provisions.

Dispute settlement Dispute settlement must not be held hostage to economic clout. Given the vast difference in economic power between Northern and Southern countries, sanctions need to be multilateral. Little is achieved if a group of small Southern countries uses sanctions against an economic giant. Also, sanctions need to be used against unilateralism.

Implementation mechanisms Structural adjustments (such as cutting tariffs and reducing perceived subsidies) are currently the implicit WTO implementing mechanisms applied to the South. However, there is no counterpart implementing mechanism to ensure that Northern countries cut their tariffs, avoid tariff escalation, reduce their subsidies or abide by WTO agreements, such as the Agreement on Textiles and Clothing. This anomaly needs to be rectified.

The above are some key examples (largely building on previous chapters) of environment-unrelated issues around which a coordinated Southern negotiation strategy could be built. From a negotiation theory perspective, packaging trade and environment along with these other issues provides us with the opportunity to 'trade across differences', since 'gains' along any of these issues could be bargained for 'concessions' on trade and environment (see Susskind 1994; Najam 1995).[6] However, before returning to 'win-win' discussions of how the developing countries may actually score 'gains' on environment and development, we will, in the next section, focus on what negotiation format best suits the larger interests of the collective South.

Towards Post-Seattle Trade Negotiation

The events of the Third Ministerial Conference of the World Trade Organization, held in Seattle, USA, in late 1999, sent a number of important messages to WTO decision-makers. Arguably, the street demonstrations by various US-based environmental and labour groups were as much a function of domestic US politics – and at least partly of the location in Seattle – as of the global trade discourse. However, the intensity of the reaction as well as the substance of the message delivered at Seattle points to a number of important messages to the developing countries. The following, in particular, are of significant concern in outlining any future negotiation strategy for the South.

1. First, there is a very clear sense in the post-Seattle world that environmental activists in the North will continue pushing ever more aggressively for the imposition of explicit environmental linkages to trade regulations.
2. Second, and more importantly, it is now evident that the groups most aggressively campaigning for environmental conditionalities are the ones who define the 'environment' most narrowly and have the least understanding of Southern concerns and realities. To put it bluntly, many of the activists in the streets were well-meaning but ill-informed about the scope and shape of developing countries' environmental challenges. The reality of Seattle was that the message of the larger and more global environmental groups, which have, over the years, developed a sensibility to Southern concerns, was trumped and overwhelmed by the more vociferous, more boisterous, and more simple mantras from the activists – mantras that sounded very much like a recipe for green protectionism.
3. Third, and finally, the developing countries were seen and portrayed as generally 'anti-environmental' and certainly as the main hurdle to the incorporation of environmental linkages into trade regimes. While the academics had been suggesting this more politely (Whalley 1996; Uimonen 1999; von Moltke 1999), the activists and the media

were far more blunt in proclaiming the South as the 'villain' in the story – the problem, as it was portrayed to and by them, is that the developing countries just don't care enough about the environment!

The challenge to the South in a post-Seattle world, therefore, is both steep and clear. First, it has to confront the reality that it is now increasingly difficult, if not impossible, to keep trade discussions de-linked from environment. Second, it has to realize that those who are now at the vanguard of the movement are the very ones who (even though they may proclaim Southern sympathies) have the least under-standing of Southern concerns. The developing countries have to invest in tutoring a whole new set of Northern environmental groups in environmental realities and priorities in the South.[7] Finally, and in conjunction with the above, the South has a major image restoration exercise before it. The developing countries have to demonstrate – through word and through action – that they are not anti-environ-mental. It is only that their environmental priorities are different from those of the Northern groups and are much more focused on issues of environmental justice, sustainable livelihoods, poverty alleviation and human security, i.e. on sustainable development.

How can the South go about meeting this challenge in post-Seattle trade negotiations? A major portion of the answer depends on the substance and structure of future international trade negotiations. In-fluencing this substance and structure should be the first priority of the developing countries between now and the next ministerial meeting. Questions about whether there should be a new 'round' of global trade talks and what should be included in and excluded from these discussions is the focus of discussions right now. The South needs to participate in this discussion vigorously and with a planned and concerted strategy.

Pre-Seattle, the South had been able to forge an internal consensus to focus on implementation issues. This consensus needs to be main-tained and strengthened post-Seattle. The argument for doing so could be reinforced further by drawing attention to the 'international treaty congestion' and the accompanying 'negotiation fatigue' that has been taking place over the last decade, particularly on issues related to the environment (Najam 2000). Negotiation fatigue has also crept into trade negotiations. Moreover, it is not without a certain irony that the period most people associate with the growth of international economic liberalization is also the period when international economic regulations have grown at an unprecedented pace. Furthermore, they have grown at a pace that developing countries have had difficulty in keeping up with, hence the need to take stock of what the South has already agreed to and to examine the state of the implementation impacts and effects.

This argument bears directly on how the South should approach any proposals for future negotiation. Although international trade bureaucrats (and key Northern countries) seem to have made up their minds on the matter, the fact remains that three main options have been on the table since before Seattle and, at least theoretically, still remain open: a) 'sector-by-sector' negotiations; b) a new integrated 'round' of negotiations where everything is linked to everything else; and c) the grouping together of 'clusters' of issues to be negotiated as distinct packages. However, a fourth, hybrid option can be constructed, which might be called a 'phased' negotiation option and which might meet the South's interests much better.

This chapter proposes that the South should call for the next round of WTO negotiations to be based on this 'phased' hybrid option, which can be derived directly from modifying the 'clustering' approach. The modifications are based on the provisos that:

- negotiations on implementation issues, 'built-in agenda' issues, Singapore issues, and 'new' issues would proceed in separate packages and in that exact order;
- linkages would exist between issues within a package and not across packages; and
- there would be an early harvest of negotiations on implementation issues, followed by an outcome of the cluster of 'built-in agenda' issues, and issues contained in the other clusters would be decided in the final phase.

This set of preferences, when overlaid on the 'standard' clustering option, gives us what we are calling the 'phased' approach.

Developing countries championing such an idea could base their case on concerns about 'negotiation fatigue' and marshal support of other like-minded delegations (including those from the North) for a 'slow-but-steady' approach to future negotiations. It should also be stressed that the phased negotiation option would create a clear hierarchy of WTO deliberations building upon the hierarchy implied in the Geneva Ministerial Declaration:

1. The highest priority would be given to monitoring the implementation of existing agreements and decisions and devising ways to keep these on track.
2. Concurrently, negotiation would begin on 'clusters' of sectors where a clear commitment to negotiate is already available from prior

negotiations. This would include mandated negotiations and reviews.

3. At a lower level of intensity, discussions may begin in working groups on identifying options and preferences in areas where there is no commitment yet for negotiation but which have been identified as possible areas for future deliberations. However, these discussions would not be considered formal negotiations until the earlier negotiations (no. 2) have been completed and progress on implementation (no. 1) is deemed satisfactory by the General Council.

4. At the lowest level, working groups may be initiated to prepare background investigation and review reports of possible new issues that may be raised by WTO members. The purpose of this exercise would be, for example, to gauge whether these issues are appropriate for consideration within the WTO.

Such an approach has the advantage that it could ultimately become the model for all future deliberations within the WTO. Conceptually, such an approach builds on the lessons of negotiation theory and would provide for a clear and principled hierarchy of deliberations that could begin simultaneously and 'mature' differentially. A clear set of criteria could determine when a particular issue would be ready for formal negotiation. Moreover, the 'incubation' period (no. 3) and 'investigation and review' (no. 4) phases would ensure that issues that reach the negotiation phase are actually 'mature' for negotiation and are likely to be negotiated relatively quickly. Thereby, it acts as a buffer against treaty congestion by ensuring that new agreements are not negotiated until prior decisions are being satisfactorily implemented.

The pros and cons of each of the four options are presented in Table 9.1. From a Southern perspective, the 'phased negotiation' emerges as the best option because a) it places the maximum emphasis on implementation of prior agreements without which no further action is warranted, and b) it makes the most efficient use of limited resources and expertise. Moreover, this remains a feasible option since it also meets a number of goals of the industrialized countries, especially by allowing for discussions to begin (in working groups) on other new issues (including, for example, trade and environment). From a negotiation theory perspective, the 'phased negotiation' option is also the most promising because it is based on a clear deliberation hierarchy and has the greatest potential for win–win bargains. For example, more meaningful action on implementation may be 'traded' for initiating working group deliberations on, say, trade and environment.

Table 9.1 Options for future WTO negotiations

Option	Pros	Cons
Sector-by-sector negotiation	Specific agreements could be reached in relatively short periods of time.	Gives unfair advantage to key industrialized countries with greatest say on agenda.
A new 'round' of negotiation	Ensures that all issues under consideration are dealt with. Theoretically this gives a veto to every country on every issue.	Tends to be very slow; requires very high level of resources and effort which places developing countries at a disadvantage.
'Clusters' of negotiations	Allows for efficient packaging of issues important to most parties within manageable time frames.	Details remain unclear. Definition of and prioritization between packages could be contentious and skewed to developed countries.
'Phased' negotiation	Provides a clear and principled hierarchy of deliberations that could begin simultaneously but 'mature' differentially.	Non-implementation of existing agreements could hold future negotiations hostage.

Why the South should take a Proactive Stand on Trade and Environment

Let us now return to the specific issue of trade and environment and how it fits into the larger strategy described above. A necessary precursor to the question we originally set out to answer is: 'Why should the South take a proactive stand on trade and environment?' In general, developing countries have been, and remain, wary of movement in the trade and environmental agenda. This fear springs from the assumption that the environment is largely a concern of the developed countries and is likely to be introduced into WTO discussions in order to placate their domestic environmental and industry lobbies in those countries. In particular, the following two concerns are generally raised:

a) Provisions in multilateral environmental agreements (MEAs) might be used to 'trump' WTO rules.
b) The imposition of environmental standards (including process standards) could become new trade barriers to thwart developing country exports.

This represents the now standard response to the issue by most LDCs (see Nath 1997; South Centre 1998; Youseff 1998). There is merit to such arguments because developed countries have exhibited a tendency to co-opt environmental concerns for what seem like protectionist purposes. Prominent examples include the tuna/dolphin and shrimp/turtle cases, the use of clean air standards to the disadvantage of Brazilian and Venezuelan refineries (all involving the USA) and the Austrian requirement to label tropical timber (see also Chapter 2). Moreover, in light of the discussion above, a valid argument is made that trade and environment cannot be the immediate priority until the large number of prior agreements are implemented in full. This becomes more disturbing since the bulk of agreements that face implementation problems from the North are exactly the ones in which the South has greatest interest (South Centre 1998).

However, as argued above, it is now quite clear that the inclusion of trade and environment linkages cannot be indefinitely postponed. This is because of the deep substantive links between the two areas and because sustainable progress on either end is dependent on the other (Brittan 1998; Sampson 1999; von Moltke 1999). If there is a lesson in the events of Seattle, it is that we are likely to see increased pressure for dealing with environmental concerns within the WTO. Moreover, we are beginning to realize that a healthy environment is as important to a healthy economy as the latter is to the former. Sooner or later, developing countries will have to confront the deep links between trade and environment (see Chapter 1). Notwithstanding the attempts by some developed countries to manoeuvre a discriminatory and protectionist capture of the environmental agenda, there are a number of compelling reasons for the South to rethink its position on this subject.

Why be the bad guy? Despite their tendency to use environmental provisions opportunistically as disguised trade barriers, many industrialized countries are even more concerned than developing countries about the implications of explicitly linking environment to trade.[8] There is a sense that at least some of these countries (particularly the USA) raise the issue primarily to placate domestic environmental lobbies. In doing so, they strategically use the expected Southern opposition as a scapegoat to deflect the blame for inaction to the South. In essence, the developing countries, particularly those leading the argument, look like the 'bad guys', even though the major industrialized countries are themselves less than eager to undertake any action in this direction. This 'politics of posturing' was ever more fervent at the Third WTO Ministerial in Seattle, where the environmental groups came prepared and created a major disruption. It would be sad if the developing countries were to be needlessly, and erroneously, seen as environmental laggards. An openness (even eagerness) to discuss a meaningful and South-friendly incorporation of trade and environmental concerns into WTO and MEAs could turn the table on these developed countries and force them to show their real hands. This would, however, require a careful analysis and clear articulation of the South's trade and environment interests.

Let's face reality There is no value in fooling ourselves about the level of clout the South has over the international trade system. The fact of the matter is that the developing countries have been consistently unsuccessful throughout GATT/WTO history in using the threat of their non-participation to influence the outcome of issues that were of importance to key industrialized countries. It should be recalled, for example, that in the 1980s the developing countries refused to accept the inclusion of issues related to services, investment and intellectual property into GATT negotiations using arguments similar to those they are now using on the environment. Yet, by the end of the Uruguay Round, each of these issues had been incorporated into the GATT. Most observers agree that on investment, as on other issues before it, the agreement reached between the OECD countries will most likely become the basis of what WTO will ultimately incorporate (see Chapter

7). The point to be highlighted is that if, indeed, the developed countries are bent on including a set of environmental clauses into WTO rules, they are likely to impose them one way or the other, sooner or later. However, if developing countries participate proactively in the discussion at this point and focus on putting forth alternative proposals on how to incorporate environmental issues into international trade regimes, they will at least have some chance of influencing the final outcome. By opting to remain 'out of the loop' at this formative juncture, they stand only to be reduced to spectators of the final outcome.

The writing on the wall   Related directly to the above, even if there is limited interest *at this point* to deal explicitly with trade and environment issues, it is obvious that this issue will have to be tackled at *some point*. A large number of multinational corporations, NGOs, academics and government agencies in the developed world are planning on this assumption and preparing to influence the shape of regulations, as and when these happen. They are already working feverishly at defining the likely future shape of trade and environmental regulations (see, for example, Whalley 1996; Brack 1998; Sampson 1999; Uimonen 1999; von Moltke 1999). By remaining out of this discussion, developing countries will only forfeit their right to influence its ultimate conclusions. For example, many companies in Europe have already adopted eco-labelling procedures. Academics and NGO-based researchers are similarly scrambling to put their mark on the emerging order (see, for example, Whalley 1996; Brack 1998; Sampson 1999; Uimonen 1999; von Moltke 1999). These ideas, experiments and pilot initiatives are already assuming a life of their own. Once these have been set in place as precedents and norms, they will influence, one way or the other, future trade rules. The ability of developing countries to have a say at that ultimate stage will be much less than it is at this formative juncture in the evolution of these regulations. The formal decisions may come later, but the real decisions are being made right now.

MEAs are our friends   In general, international environmental regimes have been more accommodating than trade agreements in providing differential, and preferential, treatment to developing countries. Arguably, developing countries can defend their interests better in fora seeking 'sustainable development' than in those advocating unadulterated trade liberalization. The urge to retain trade fora (where they have traditionally been at a disadvantage) as the ultimate arbiter over MEA provisions could backfire. For very understandable reasons, developing country governments tend to take trade negotiations much more seriously than environmental negotiations. Yet the track record suggests that developing countries get more sympathetic hearings at MEA forums. Needlessly antagonizing the environmental community will serve little purpose and could end up having strategic long-term costs. Moreover, one must note that the standard norm of international law is that the more specialized treaty will take precedence on any given issue unless specifically stipulated otherwise (the principle of *lex specialis*), which means that MEAs will generally trump trade law on issues related to the environment.

Whether governments engage or not, exporters will In a very practical way, the decision on whether or not to engage in the trade and environment debate is really beyond the control of Southern governments. International markets in general and Southern producers (via their Northern customers) are already making the decision for them. There is a strong trend among export producers in the South to move towards environmental measures – not out of fears of future WTO rules, but because their markets and consumers in the North are demanding it. The real change may be happening not because of the WTO, but despite the WTO. This trend within the market – for example, the leather industry in Pakistan – is far more real than anything that happens within WTO negotiating rooms. Focusing only on WTO and merely on blocking all and any mention of the environment in that forum may turn out to be not only a losing battle, but a meaningless battle. The much more important battle for the Southern governments is to ensure that the environmental steps that their producers take are not taken out of duress. The focus should be on ensuring that these environmental measures undertaken by export producers are not merely intended to placate environmental consumers in the North but genuinely contribute to the pursuit of sustainable development in the South.

What is good for the South's environment is good for the South's development Finally, let it not be forgotten that if something is good for the South's own environment then developing countries should aggressively pursue it irrespective of what the North may feel about it. Just as it would be a folly to be bullied into an agreement that does not serve the South's developmental and environmental interests, it would be even sadder to refuse options that *do* benefit the South's own environment and populations just to spite the North. There are many possible trade and environment provisions that would benefit not just the South's environment but also their economy.[9] The developing countries should focus on these. The opportunity is to convert the 'trade and environment' debate into a 'trade and sustainable development' debate and the challenge is to accentuate the opportunities that it provides while deflecting the threats.

Towards a Positive and Proactive Trade and Environment Strategy for the South

The obvious conclusion from the above is the urgent need to shift Southern strategy on these issues from negotiating 'reactively' to negotiating 'proactively'. Instead of merely reacting to the agenda set by the North and then complaining about how it does not meet Southern interests, the developing countries have to invest in defining and defending an agenda that meets their own interests. Given that the ability to define the agenda translates directly into negotiative power, developing countries need to move beyond why they find the proposals made by the developed countries unacceptable to focus on defining alternative proposals that do meet their interests (Najam 1995, 1998).

While defining the contours of such a position would obviously require systematic intellectual effort and intensive negotiations among developing countries, some general elements of such a position can be identified here as a preliminary exercise.

A focus on sustainable development Any discussion of trade and environment within the WTO should be in the context of 'sustainable development' as defined and discussed in *Agenda 21* (emerging from the 1992 Rio Earth Summit) (UNCED 1993). This would include the focus on differentiated responsibility for developing countries, adherence to the right to development, and recognition that international environmental obligations undertaken by the developing countries are subject to the provision of adequate multilateral assistance (Youssef 1998). Both these principles are repeated in most MEAs and should be re-articulated and operationalized within any WTO provision.

Just say NO to protectionism The WTO should clearly articulate the principle that the environment should not be used as a protectionist barrier to trade. This would include clear rules against the use of eco-labelling or other exclusionary devices based on unrelated process and production methods (PPMs) and transparent and accessible mechanisms for addressing disputes arising from an interpretation of what is or is not protectionism.

Subsidiarity rules All trade and environment discussions should recognize the principle of subsidiarity – that priority should be assigned to the lowest jurisdictional level of action consistent with effectiveness. International policies should be adopted only when these are more effective than policy action by individual countries or jurisdictions within countries.

Taking small steps towards MEA–WTO coherence The goal should be to achieve substantive and jurisdictional coherence between the WTO and MEAs on all issues related to trade and environment. Moreover, WTO as well as MEA compliance should be understood in the context in which these agreements were negotiated. This would mean, for example, that any action would only be relevant if all parties are full members of the WTO as well as of the MEA in question. More specifically, this would also mean that violation of MEA requirements due to the lack of international contextual conditions (such as international assistance) should not be a subject of WTO action.

Pooling environmental provisions within the WTO All trade and environment issues should be dealt with together by pooling the various environmental provisions now scattered within different WTO provisions. Currently, WTO agreements refer explicitly to the links between trade and environment in four instances: Article XX of general exceptions in GATT 1994, the agreement on technical barriers to trade, the agreement on sanitary and phyto-sanitary measures, and the agreement on trade-related intellectual property rights (TRIPs). These and any other environmental provisions should be negotiated as a package, including clear rules about the patenting of life forms (as in the 'Texmati' case)[10] and trade in genetically modified organisms (GMOs).

Avoiding trade restrictive measures The application of trade-restrictive measures should be a device of last resort. It should be applied only when all other means of improving MEA compliance and environmental conditions have been exhausted. In particular, MEAs should ensure the provision of technical and financial assistance to developing countries to facilitate conversion to environment-friendly processes and methods and the availability of relevant technology that can be absorbed and adapted by developing countries.

Conclusion

Table 9.2 provides a summary of some of the key points raised in this chapter. In essence, the position adopted by developing countries

Table 9.2 The case for adopting a proactive stand on the environment

Why the South should worry about trade and environment	Why the South should take a proactive stand on trade and environment	Elements of a proactive negotiating agenda for the South
• Environment might be used to 'trump' WTO rules. • Environmental standards can be used as trade barriers and as a guise for protectionist policies. • Environmental concerns can distract from the South's more pressing and legitimate developmental priorities.	• Developing countries are likely to be used as scapegoats. • They will lose the chance to influence the emerging shape of the trade and environment debate. • MEAs are more accommodating in providing differential, and preferential, treatment to developing countries. • LDCs can still remain vigilant on their concerns while pursuing a proactive agenda. • Irrespective of WTO rules, Southern export producers will have to respond to environmental concerns because of consumer demands. • What is good for the South's environment has to be good for the South.	• Sustainable development and *Agenda 21* provisions. • Principle of not using environment as a trade barrier. • Principle of subsidiarity. • Coherence between MEA and WTO provisions and compliance. • Trade restrictive measures should be device of last resort. • Deal with all trade and environment issues together.

until now has consisted of trying to stall any movement on trade and environment issues in the WTO. Although the motivations for doing so are understandable, this (non-)strategy has backfired. On the one hand, it has allowed industrialized countries to use developing countries as scapegoats. On the other hand, the developing countries have effectively removed themselves from a position where they could influence the emerging discussion on the subject and steer it in a relatively South-friendly direction. This chapter concludes that this strategy should be abandoned and a more positive and proactive approach adopted immediately.

The message of this chapter is blunt and simple. Whether the South likes it or not, environmental concerns *will* be an important determinant of future trade regulation. If the developing countries are actively engaged in this discussion, there is at least the possibility that environment will be dealt with in a manner that does not retard, and may even advance, the developmental aspirations of the South (under the rubric of sustainable development). If the South chooses to disengage itself from this discussion (as it has on other issues),[11] then the probability is that the environment will, in fact, be used as a protectionist device that will only relegate the South's priority issues to an even lower footing than in the prevailing trade regime.

The choice before the developing countries is stark. The easy option is to opt out of the trade and environment debate on the (legitimate) grounds that it is driven by narrow Northern agendas and is potentially a guise for green protectionism. The more difficult, but clearly more advisable, option is to become actively engaged in the discussion and change the terms of the debate by injecting a 'Southern environmental agenda' into the discourse – one that places the trade and environment discourse within the broader framework of sustainable livelihoods, poverty alleviation, social and environmental justice, and the expansion of ecological and human security, i.e. of sustainable development. One cannot, and should not, expect such a Southern agenda to come from the North – the onus for defining it lies squarely with Southern governments, scholars and policy NGOs.

The recipe for the second, and preferred, option includes a concerted and focused agenda of policy research and analysis so that the developing countries can better articulate their own environmental priorities and their linkages to international trade. From a negotiation perspective, it will require intense negotiations not just with the North but within the South. It will mean building bridges between North and South and building coalitions with like-minded countries and groups. Most importantly, it will mean that the developing countries have to move from a 'reactive' negotiation strategy (where they are essentially reacting to an agenda set by the industrialized countries) to a 'proactive' negotiation strategy where they themselves put alternative agendas – such as the one proposed here – on the table.

Notes

1. Assistant professor of international relations and environmental policy, Boston University, USA, and visiting fellow, SDPI. The author gratefully acknowledges the assistance and advice of Shahrukh Rafi Khan.

2. These concerns are not repeated here because they have been so lucidly articulated elsewhere in this book (see also South Centre 1998). Furthermore, the main Southern argument is by now well recognized and paraphrased in the larger literature, including that emanating from the North (for example, Runnalls 1996; Brack 1998; von Moltke 1999).

3. For a much more detailed discussion of the concept of the 'South' and the rationale and mechanisms for increased Southern coordination see Najam 1995, 1998. It is important to note that the goal of a coordinated Southern negotiation strategy is not as much to 'oppose' the North as to actualize the common goals that bring the coalition together in the first place.

4. The following points are based on text provided by Shahrukh Rafi Khan.

5. On issue (c), careful research shows that high rates of protection generated inefficiency and hence adversely affected growth (see Little, Scitovsky and Scott 1970). However, the real issue is an 'internal competition policy' and not protection. If countries have a competition policy (deregulation and monopoly control), they do not need trade liberalization to achieve competition, i.e. trade liberalization is not the first best policy. Countries such as Korea retained protection but ensured internal competition.

6. For example, developing countries could trade concessions on the environment for major and meaningful liberalization on agriculture, textiles and clothing, leather and other sectors of their interests.

7. The challenge to the global groups (such as the IISD, the IUCN and the WWF) and their Southern partners that had, until now, been investing in bringing Southern governments and NGOs on board the trade and environment band-wagon has also changed. Under post-Seattle realities, the much more important challenge for these groups is to educate Northern environmental groups on the real environmental and developmental concerns of the South, to establish North–South civil society dialogue, and to 'de-demonize' the role of developing countries in international trade and environmental discussions.

8. This has been abundantly obvious, for example, in the negotiations on bio-safety (under the Biodiversity Convention), where the US position has been very similar to the Southern position at the WTO: 'No MEA should be allowed to trump WTO rules.'

9. This includes not just the fact that a better environment can mean less poverty and, therefore, a better quality of life, but also more tangible issues such as access to 'niche' markets and the creation of advantages through aggressive certifications under the ISO 14000 series, and so on.

10. The 'Texmati' case refers to the WTO dispute between USA on the one hand and Pakistan and India on the other regarding the US patenting of a Texas-based rice variety under the name 'Texmati'. The Texmati variety has been developed to be essentially similar to the famous Basmati variety traditionally coming from India and Pakistan and capitalizes on its premium market value by using a similar name.

11. For example, on the issue of global climate change the developing countries have essentially accepted an agenda defined by the North and, for all effective purpose, have opted out of the main negotiation process simply because they are not currently obligated to reduce emissions. As a result, the South has chosen to relegate itself to the role of a silent spectator (or, at best, merely a 'sink' for emissions trading) in most debates about emission reduction obligations. What the developing countries seem to neglect is that the rules now being framed for emissions reduction – with little to no serious input from them – although not currently applicable to them will, eventually, also apply to them (See Najam 1997; Najam and Sagar 1998 and Chapter 6).

Bibliography

Bilginsoy, C. and S. R. Khan (1994), 'Cross-sector export externalities in development', *Economic Letters*, Vol. 44: 215–20.

Brack, D. (ed.) (1998), *Trade and Environment: Conflict or Compatibility*, London: Earthscan and the Royal Institute of International Affairs.

Brittan, L. (1998), 'Trade and environment after Singapore', in D. Brack (ed.), *Trade and Environment: Conflict or Compatibility*, London: Earthscan and the Royal Institute of International Affairs.

Jha, V., G. Hewison and M. Underhill (eds) (1997), *Trade, Environment and Sustainable Development: A South Asian Perspective*, London: Macmillan.

Little, I., T. Scitovsky and M. Scott (1970), *Industry and Trade in Some Developing Countries*, London: Oxford University Press.

Najam, A. (1995), 'International environmental negotiation: a strategy for the South', *International Environmental Affairs*, Vol. 7, No. 2, 249–87.

— (1997), 'Global climate change negotiations – how should the developing countries respond?', *South Letter*, Vol. 29, Nos 3/4: 14–16.

— (1998), *International Environmental Negotiation: The Case for a South Secretariat*, SDPI Monograph Series, Monograph No. 6, Islamabad: Sustainable Development Policy Institute.

Najam, A. (2000), 'The case for a law of the atmosphere', *Atmospheric Environment*, Vol. 34, No. 23: 4047–9.

Najam, A. and A. Sagar (1998), 'Avoiding a COP-out: moving towards systematic decision-making under the climate convention', *Climatic Change*, Vol. 39, No. 4.

Nath, K. (1997), 'Trade, environment and sustainable development', in V. Jha, G. Hewison and M. Underhill (eds), *Trade, Environment and Sustainable Development: A South Asian Perspective*, London: Macmillan, pp. 15–20.

Raghavan, C. (1999), 'The new issues and developing countries', Third World Network Trade and Development Series, http://www.twnside.org.sg/souths/twn/title/rag-cn.htm.

Runnalls, D. (1996), 'Shall we dance? What the North needs to do to fully engage the South in the trade and sustainable development debate', http://iisd.ca/trade/dance.htm.

Sampson, G. P. (1999), 'Trade, the environment, and the WTO: a policy agenda', ODC Policy Paper, Washington, DC: Overseas Development Council.

Shahin, M. (1998), 'Developing country perspective', in D. Brack (ed.), *Trade and Environment: Conflict or Compatibility?*, London: Earthscan and the Royal Institute of International Affairs, pp. 150–63.

South Centre (1998), *The WTO Multilateral Trade Agenda and the South*, Geneva: South Centre.

Susskind, L. E. (1994), *Environmental Diplomacy: Negotiating More Effective Global Agreements*, New York: Oxford University Press.

Uimonen, P. (1999), 'The environmental dilemmas of the World Trade Organization', in J. J. Schott (ed.), *Launching New Global Trade Talks: An Action Agenda*, Special Report 12, Washington, DC: Institute for International Economics.

UNCED (1993), *Agenda 21: Programme of Action for Sustainable Development*, New York: United Nations.

Von Moltke, K. (1999), 'Trade and the environment: the linkages and the politics', paper presented at Roundtable on Trade and Environment, Canberra, 25 August 1999.

Whalley, J. (1996), 'Trade and environment, the WTO, and the developing countries', in R. Z. Lawrence, D. Robrik and J. Whalley (eds), *Emerging Agenda for Global Trade: High Stakes for Developing Countries*, Policy Essay No. 20, Washington, DC: Overseas Development Council.

Youssef, H. (1999), 'Special and differential treatment for developing countries in the WTO', Trade Working Paper No. 2, Geneva: South Centre.

. . . .

CHAPTER 10

Summary and Concluding Considerations

Shahrukh Rafi Khan

While individual authors have included summaries and or conclusions at the end of their chapters, it is often useful to get the gist of a book in one place. I have also used the editorial prerogative to include some concluding considerations at the end of this chapter.

In Chapter 1, Aaron Cosbey examines the complex linkages between trade and the environment. He reviews how trade creates wealth via allocative efficiency, competition and the use of more efficient imported machinery and embodied methods. However, negative environmental impacts can also result from trade. As scale expands, so do accompanying environmental problems, particularly if existing regulations are lax or implemented poorly. More wealth can lead to more wasteful consumption. Standards may be kept inappropriately low for strategic export gain. However, adopting appropriate standards, possibly because of importer pressure, may result in winning niche markets and reduced input use and pollution.

Cosbey does not discount the existence of 'green protectionism' or 'eco-imperialism'. He points out that sometimes standards can be tailored to a particular sector that is threatened by Southern imports. Eco-imperialism refers to standards that are *process*- rather than *product*-specific. The former are particularly odious for the South, since they prefer to make their own decisions regarding how to produce. Both kinds of standard, if imposed, can deny the South the benefits of trade and hinder prospects of North–South collaboration.

In Chapter 2, Shahrukh Rafi Khan reviews the trade and environmental literature that deals with a number of issues and hypotheses that are not a part of traditional trade theory. Many of these are related to concerns in the North or the South about fair trade. First, trade liberalization could result in strategic movement on the part of Northern multinational corporations to Southern countries with more lax environmental regulations and hence result in a loss in Northern jobs. Second, the North could use trade liberalization to dump its dirty technology and other domestically prohibited goods (DPG) on the South. Third, structural adjustment-induced export promotion could result in the South exporting its environmental capital in the form of high pollution and domestic resource degradation. Fourth, the multilateral environmental agreements (MEAs) are increasingly affecting the world trading environment and these MEAs could block Southern exports. Fifth, the North has a greater resource and technological ability to meet the standards it sets and this will mean blocking access to Southern exports and enhancing its market share. Sixth, the cost of mitigating such pollution in the South is very high.

The next two chapters examine the empirical content of some of these propositions. In Chapter 3, Tariq Banuri reviews the cotton commodity chain from a Pakistani perspective. A key theme of the chapter is governance by 'markets' and 'government' at the various stages in the cotton production chain. Market-driven governance systems are shown to be more flexible. Banuri also gives a full account of the trade-related environmental impact of production at the various stages of the commodity chain and presents environmentally sustainable alternatives.

Cotton and cotton-based products are the main export of Pakistan. In one form or another, about 70 per cent of cotton gets exported from Pakistan, and the textile sector accounts for 64 per cent of exports. This poses a serious environmental problem due to the nature of cotton production and processing. The main obstacles to the expansion of cotton yields have been the inadequacy of water and pest attacks. Addressing these problems has been the main source of environmental degradation. The establishment of irrigation systems has resulted in rising water tables, waterlogging and salinity in an increasingly water-scarce country. About 65 per cent of the total pesticides use in the country is for cotton and between 95 and 98 per cent of cotton-growing area in the Punjab province is treated with pesticides. Volumes in use are still less than those of the high-pesticide-use countries. For example, Central American producers spray between 20 to 30 times a season compared to 8 to 13 times in Pakistan. Nevertheless, severe environmental impacts are already evident.

Pesticides affect human health as well as wildlife and domestic animals, biological diversity, surface and ground water quality and fish population, and erode soil quality and fertility. Other sources of contamination include run-offs into water bodies, improper use of pesticide containers and inadequate or illegal disposal of expired and unused pesticide. Pesticide poisoning results in various ailments including stomach cramps, dizziness, vomiting and heavy sweating. Countries such as Pakistan are on a 'pesticide treadmill' such that higher and higher doses are needed to contain pests as they develop immunity (meanwhile, friendly predators are wiped out). Higher demand leads to higher prices and costs of production. Even so, there have been dramatic outbreaks of pest attacks.

There is a strong economic case for eliminative pesticides, since alternatives such as organic cultivation and integrated pest management (IPM) are slow in gaining acceptance. The transformation of cotton production would not represent an economic problem since it constitutes only 10 per cent of the cost of the finished garments. Also, there is evidence that green cotton products could command up to a 20 per cent mark-up. The real constraint lies in the domain of governance and technology transfer rather than production costs and consumer preference.

The main problem of conversion is that it requires collective action for the whole sector and, given the lack of information, short-term losses and uncertainty, individual farmers do not have an incentive to convert, unlike in the case of the Green Revolution. While the government has an incentive to maximize aggregate output in the long run, its ability to bring about the needed transformation has eroded over time. The research and extension infrastructure it put into place for the Green Revolution is suffering because of low salaries, bad work conditions and low morale. Poor monitoring has also led to a deterioration of efficiency.

Meanwhile, the market represents an alternative governance mechanism that is manifested via highly concentrated multinational firms that have captured parts of the input market such as seed and insecticide supply. They have the incentive to maximize profits and in this regard are highly nimble in making adjustments as needed, unlike the state sector. The upshot is that they aggressively market and create a dependence on pesticides, and the monopolistic market structure results in imposing a heavy burden on farmers. Given this scenario, moving to alternative cotton cultivation represents a heavy challenge for the government in ensuring that the right incentives exist across the board. A roundtable process involving all stakeholders is recommended for initiating the process.

The other two parts of the global commodity chain include yarn and cloth production and apparel manufacture. While apparel manufacture is subject to the governance structure of mass retail firms, yarn and cloth production do not exhibit a coherent governance structure. The textile industry is subdivided into spinning, weaving, processing and made-up products. Cotton processing is highly polluting since it utilizes a number of chemicals. The main source of pollution is the discharge of untreated effluents into water bodies and soils. The effluents contain organic and inorganic chemicals as well as suspended solids (fabric and grease). By lessening the dissolved oxygen in water bodies, the effluents threaten aquatic life.

The textile-processing sector, which is of greatest concern environmentally, consists of large professionally organized firms. There is a good chance that a sensible governance system will emerge. The garment and apparel sector is tied via the global commodity chain to international retail chains that impose a governance system via specification of labour and other standards for suppliers. A model of sustainable industrial production involving a roundtable process of all stakeholders and self-monitored standards by industry is described.

The export market of finished textiles has been jolted in recent years by health- and environment-related trade restrictions. Thus European countries have imposed bans on products made with azo dyes, and corporate retail firms are able to impose production standards on small and decentralized buyers. The monopolistic market structure also means that large retail firms capture most of the rents. Surplus in the chains results from changing consumer preferences and they accrue at the higher ends of the chains (such as mass retailers), which are located in the North. Thus only 9.4 per cent of the total value of finished cloth accrues to the farmers, whereas the bulk of the rents accrue beyond the finishing stage in the production of garments and retail to customers in the North. Such a distribution results from the market power of large US and European retailers.

In this regard, as earlier indicated, raising the price of cotton by moving to organic cotton production would not matter much from an overall market/trade perspective. However, this option is not straightforward from a production perspective, since the market, as yet, is quite limited and the premium not uniformly high. Banuri suggests that IPM (integrated pest management) has more scope in Pakistan but little real government backing and that, as yet, the prospects of higher yield are uncertain. Partly for this reason and partly due to inertia, the lack of information, and corporate lobbying, the government agricultural machinery still backs the chemical-intensive 'Green Revolution' production methods.

Shahrukh Rafi Khan et al. take the empirical analysis a step further in Chapter 4 by computing the costs and benefits of pollution mitigation in cloth production and leather tanning. One of the important propositions of the trade and environment literature is that exports will generate a great deal of pollution in poor countries because of the environmentally unsound production methods they utilize. Another proposition in this literature is that the costs of reducing such pollution in poor countries are very high. Both these propositions are explored using data from cloth production and leather tanning, two of Pakistan's strategically most important but very dirty production processes.

The increase in Pakistani exports of cloth and leather and footwear that could result from the reduction in trade barriers agreed to in the Uruguay Round in 1994 is forecast based on past trends and work done by others. Between 1996 and the end of 2004, cloth exports from Pakistan can be expected to rise by 45 per cent and the corresponding increase in pollution load is calculated to be 81 per cent. Leather exports are expected to decline, so one could expect a 7 per cent lower pollution load generated by leather tanning. If the suggested measures are adopted, up to 91 per cent of the emissions from cloth and 66 per cent of the emissions from tanning could be reduced.

The costs of such measures in cloth production, for the economy as a whole, would have been 0.12 per cent of GNP in 1996. The foreign exchange liability, in terms of imported equipment, for this year would have amounted to 1.6 per cent of only cloth exports in 1996. More important, the cost to industrialists from adopting measures to reduce pollution would have been 1.6 per cent of their sales revenue. The costs of such measures in leather production, for the economy as a whole, would have been 0.0048 per cent of GDP and the mitigation cost to exporters of leather would have been 0.88 per cent of their export revenue. These mitigation costs are much lower than for cloth production since clean production technology is locally available.

Thus the evidence from the case study supports the first proposition that export growth may generate a great deal of pollution. However, there is little support for the second proposition that the costs of establishing and operating clean technology in the South are necessarily very high. Also, a rough calculation of the benefits indicate that these are 0.5 per cent and 0.04 per cent of GDP for textile and clothing and leather respectively, and these far exceed the costs. Admittedly, the estimates are crude, but the benefits exceed the costs by such a large margin that it is unlikely that refinements would change the main message that, from society's perspective, it would pay to encourage the adoption of standards for reducing pollution.

In Chapter 5 Haroon Ayub Khan and Abdul Matin Khan describe the Pakistani experience of the response to environmental standards. The key aspect of the response was the revision of the Environment Protection Ordinance 1983 in the form of the Environment Protection

Act 1997. The new Act specified a pollution charge, among other punitive measures, for the violation of the National Environment Quality Standards (NEQS) for air and liquid emissions, on which work was already under way. Given the limited capacity of the National and Provincial Environmental Protection Agencies to implement the Act in general and the NEQS in particular, the government engaged in a consultative process with business and civil society to find an appropriate solution. The key element in the solution was self-monitoring and reporting by industry regarding their compliance with the NEQS and the voluntary payment of the pollution charge in case of violation. The remarkable achievement was that industry itself came up with the base pollution charge and an associated schedule of graduated rates. The thinking embedded in the rate schedule is that the charge should be a 'slap on the wrist' in the first year and should 'draw blood' in year five so that it is cheaper to comply than to pay the fine.

Khan and Khan point out that there are some important lessons one can draw from the Pakistani experience. First, intellectual leadership and mobilizing concept champions at a high level in government, industry and civil society are central to the success of the process. Second, the consultative roundtable process with government, business and civil society worked very well. The institutionalization of a continued role for civil society is important. Thus, for example, all monitoring reports are to be kept in the public domain and a role for credible environmental NGOs in each stage of the monitoring process has been institutionalized. Third, the leadership of key members of civil society in facilitating the roundtable process in a neutral manner has been crucial. Also, when needed, NGOs have applied pressure to keep the process moving and on track. Fourth, the technical capacity of the EPAs for monitoring compliance needs to be built concomitantly with the negotiation process. The fact that this capacity-building has lagged behind has resulted in implementation delays and credibility problems. Fifth, training, capacity-building and awareness-raising in industry is also necessary right from the inception of the process. Despite the problems – some expected foot-dragging on the part of industry, bureaucratic inertia and genuine unresolved issues, such as who will collect the pollution charges and how they are to be used – substantial progress has been made, with the pilot phase well under way in early 2001 and implementation planned for 1 July 2001.

The third section of the book contains emerging issues on the horizon regarding the trade and environment interface. Chapter 6, by Aaron Cosbey and Victoria Kellett, examines the trade and environmental implications of the climate change negotiations currently under way. In December 1997 in Kyoto, Japan, developed country parties (Annex I countries) agreed to targets to decrease overall emission of six greenhouse gases (GHG) to 5.2 per cent below 1990 levels between 2008 and 2012. However, they also negotiated three mechanisms via which to increase the flexibility available to them in meeting targets. These included international emissions trading within Annex I countries, joint implementation within Annex I countries and project-based clean development mechanisms (CDM) for investment in non-Annex I countries that contribute to sustainable development (SD) and decrease GHG.

CDMs, which are at the trade–environment interface, will be quite complex to implement. For example, there is a possibility that certified emissions reduction (CER) credits will be traded in the open market. This is a mechanism of using trade, in a strictly market sense, to achieve environmental improvement. However, it is also a mechanism that Southern critics argue enables the North to preserve its environmentally destructive lifestyle. CERs emerge from projects implemented by a Northern foreign investor and a developing host country to bring about sustainable development and GHG emission reductions. The credits are to be shared for a CER-producing project by the investing and host country. This thus represents an innovative mechanism in which trade can achieve environmental improvement.

The range of projects that contribute to SD, such as investment in renewable clean energy, transmission losses and possibly forestation, are manifold and very promising. Thus the potential for enhancing SD via financial and technological inflows and capacity-building, rural development and energy self-sufficiency are in principle welcome.

Konrad von Molkte in Chapter 7 examines the need for an international investment regime in a broader sense. For many countries in the South, foreign direct investment (FDI) has become the most important source of resource inflow, overtaking both capital inflows and funds made available by multilateral development agencies. Between a third and a half of private corporate investment is undertaken by affiliates of foreign corporations. An investment regime consistent with SD is unlikely to be a simple extension of the GATT/WTO trade regime, as is being contemplated, for various reasons. In particular, the static nature and insensitivity to SD issues, such as the environment, which GATT/WTO views as a trade barrier, would rule this out. Thus, given the static character of the existing trade regime, it may well collapse if an investment regime is appended with the same institutional characteristics as the existing trade regime.

Von Molkte views the principles of an investment regime from the perspective of sustainable development. The interpretation of the term 'like' products is likely to present the most serious conflict between the GATT/WTO and the requirements of SD, which entail distinguishing between products produced sustainably from the rest. The GATT/WTO has continued to resist doing this in its understanding of 'like'. This interpretation of like is carrying over to the framing of investment agreements that refer to 'in like circumstances'.

Unlike trade in goods, productive investments create a complex system of rights and obligations extending into an indefinite future. Not recognizing this broadened social dimension could therefore contribute to the destruction of social and environmental values. Given the long-term nature of investment, an investment regime would thus need to address this temporal nature of investment and recognize that 'like circumstances' can involve variation as circumstances change.

Because of the complex and imprecise nature of environmental management, including regulations, incentives, informational obligation overlaid with changing technologies and the 'absorptive capacity' of the environment, 'like circumstances' will differ for the same facility over time. Given this dynamic and complex process of effective environmental management (EM), the institutional capabilities of an investment regime in determining what 'national treatment' is will be seriously challenged (ironically, judging from NAFTA, foreign investors have rights not available to domestic counterparts). Given the complexity of the EM process, dispute settlement procedures need to be framed with great caution. The principles of a trade regime such as the WTO/GATT are thus not appropriate for an investment regime, and the differences need naturally be reflected in a dispute settlement process. Similarly, the needs of environment and sustainable development may require a more nuanced interpretation of the most favoured nation (MFN) principle. Companies from countries with a poor track record on environment management may rightly require greater scrutiny.

The issue of 'investor' versus 'host' country responsibility is also viewed as a fundamental principle of an investment regime that can have an important impact on the environment and sustainable development. An important question that arises in this regard is what standards of enforcement should an investor be held to – national or the more rigorous standards of the investor country? Von Molkte's sensible response is that this may depend on the nature of the investment. For example, if the investment is a component that feeds into a global commodity chain, the investment should be held accountable to the more rigorous standard. Similarly, where impacts are cross-border, i.e., pertaining to the stratosphere, climate change or sometimes biodiversity, the more rigorous standards should apply.

Foreign investment is discussed as part of the Multilateral Agreement on Investment (MAI). While that initiative did not succeed, the issue is unlikely to go away. For this reason, and also because investment can have a far-reaching impact on sustainable development in general and on the use and protection on natural resources in particular, it remains an important issue. Alternatives to non-sustainable development in the South often require investment. In principle, the South should welcome an investment regime that makes the risks more calculable and hence reduces the rate of return needed for the flow of funds to the South.

Investment agreements have either looked at the issue from the perspective of investors (Northern perspective) or the rights of the host country (Southern perspective), with little attempt made to draw these perspectives together. Launching the MAI as a closed technical exercise without a mandate was therefore an error.

To sum up, von Molkte considers the basic principles of a trade-related investment regime from the perspective of sustainable development, and presents an alternative framework agreement on investment that would retain both predictability and flexibility. He concludes that given the particular nature of investment, relative to the movement of goods and services, this would require an agreement outside of the existing international organizations including the WTO.

The two chapters of the concluding section of the volume provide an agenda for the South for the trade and environment interface. Mark Halle in Chapter 8 provides a practical road-map for Southern countries. His main point is that the South will inevitably have to deal with the issues pertaining to the trade–environment interface. Given that, a judicious use of trade rules can advance their interest. Furthermore, Halle suggests ways to move Southern countries rapidly, effectively and affordably to a point where they can follow and participate in the debate on trade and the environment, analyse and articulate its legitimate national interests and advance these interests successfully in the international arena where such issues are debated and major decisions taken. In this regard, suggestions are made regarding information, networking, research, capacity development, institutional arrangements, transparency and regional cooperation.

Adil Najam's concluding chapter provides practical suggestions to the South for dealing with future WTO negotiations. He starts by presenting a Southern perspective on trade issues. Following this, he takes up options for advancing the WTO negotiations in a way that addresses Southern needs, and explains why it makes sense to adopt a positive and proactive negotiating agenda and what the key elements of such an agenda are. The point that stands out in particular and bears repeating is that the South should be proactive and have a positive agenda on the environment if they want to have a say in shaping the negotiations in the various ways suggested. By staying out of the discussions, they will eventually have to accept the outcome anyway and, in the meantime, will also have to take the flak for being reticent. This will happen even as several Northern countries, including the USA, are adopting similar positions to Southern countries for different reasons (such as with reference to the bio-safety protocol). Thus environmental activists in Seattle castigated Southern countries as the villains, and the media picked this up.

Najam makes several interesting observations regarding the trade–environment interface. First, the South could trade concessions on the environment for major and meaningful liberalization on agriculture, textiles and clothing, leather and other sectors of interest. Second, environmental or sustainable development fora have been more accommodating to the special and differential treatment sought by the South and should be viewed as an opportunity, not a threat. Third, and a recurring theme of the book, heeding their environment is in the South's own interest.

Trade is viewed as almost sacred by liberal economists and its presence has certainly been taken for granted in this volume so far. As an economic activity, it dates back to the inception of civilization. However, the powerful intellectual case for aid was developed by the classical economist Ricardo, who established that trade could make both parties to the transaction better off based on their comparative advantage and hence improve world welfare. This analytical exercise demonstrating the gains from trade was done in a 'static' setting, based on the comparative advantage as it exists at a point in time.

Newly industrialized countries were not impressed by the logic of comparative advantage and were unwilling to let their current and future pattern of production be determined by their current comparative advantage in resources and knowledge. In fact, they argued that the present would lock countries into their current backward state and that the state needs to be active in changing the nature of comparative advantage by strategic investments in human capital and 'sunrise' or growth industries. This obvious insight was vindicated by their development experience, although economic liberals refuse to accept that the state could ever second-guess the market. But then, ideology tends to put blinkers on those willing to be ruled by it.

For many development scholars, the notion of dynamic comparative advantage was an important challenge to existing static trade theory. An equally significant challenge is posed by environmentalists to static trade theory. In fact, this is partly based on a much broader challenge and one very early acknowledged by mainstream economics. This is the problem of the negative or positive impacts of economic activities that are not taken into account by market prices. Thus, based on the incentives these prices generate, too many or too few of these activities may take place from societies' point of view. These impacts are referred to as negative or positive externalities in economic jargon, and trade can exacerbate such impacts.

An example is the negative impacts that may result from the export of timber. Deforestation may result in soil degradation and the silting up of waterways and dams. Further, it could result in the loss of biodiversity. The market prices that exporters respond to do not reflect these costs imposed on society and so they may export much more timber than is optimal from society's point of view, particularly when one takes into account the loss to future generations. Trade can thus exaggerate the negative externalities resulting from economic activities as the market size is enhanced due to trade. This has been referred to in Chapter 1 as the scale effect.

A good example of these negative externalities is the environmental impact of the North American Free Trade Agreement (NAFTA) trade liberalization of corn production in Mexico. Nadal (2000) points to the migration from rural areas and the subsequent loss of traditional knowledge resulting from liberalization and trade-induced specialization and monoculture. The latter also resulted in soil erosion due to intensive agriculture and the negative environmental impacts of chemical fertilizer use. The pressure on land, aquifers and forests has been documented. As such examples accumulate due to the neo-liberal push for trade liberalization, countries will need to pause and seriously review the negative environmental impacts, particularly since the benefits of trade are more likely to accrue to rich business interests and the costs to the poor, who depend on the environment for their livelihoods, are the most vulnerable to its degradation and also the least capable of protecting themselves from the dirty water and air that environmental depredations result in.

There is yet another environmental reason why perhaps trade should no longer be taken for granted. Trade is premised on the continued availability of cheap fossil fuels. Thus trade has a severe negative impact on the global environment, since it requires the carrying of goods, often to distant foreign markets premised on the liberal free-trade economic philosophy that underpins this aspect of globalization. It is quite likely that the era of cheap energy will soon come to an end. Thus economic managers in poor countries would do well to pause in viewing the globalization agenda of the international financial institutions as the only game in town and consider alternatives such as the labour-intensive sustainable livelihoods approach proposed by the Society of International Development and others (see Amalric 1998).

Robins and Roberts (2000: 24) reinforce the above point by suggesting that international trade entails 'a range of community and environmental problems associated with land, sea and air freight transportation'. Further, they point out that 'much of the debate on trade and environment significantly underplays the direct and indirect impacts of trade itself, instead mostly focusing on production practices'. They cite a WTO study that concludes that 'trade as such is rarely the root cause of environmental degradation, except for the pollution associated with the transportation of goods'.

Bibliography

Amalric, F. (1998), *The Sustainable Livelihoods Approach: General Approach of the Sustainable Livelihoods Project 1995–1998*, Rome: Society for International Development.

Nadal, A. (2000), *The Environmental & Social Impacts of Economic Liberalization on Corn Production in Mexico*, Gland: WWF and Oxfam.

Robins, N. and S. Roberts (eds) (2000), *The Reality of Sustainable Trade*, London: International Institute of Environment and Development.

Index

Zed Titles on International Trade Issues

With the November 2001 Doha ministerial meeting of the World Trade Organization launching yet another round of international negotiations, trade issues will continue to dominate the world economic agenda. Zed Books has published a select number of titles which are crucial to understanding the issues involved.

Already Available

Christian Comeliau, *The Impasse of Modernity: Debating the Future of the Global Market Economy*

Carlos M. Correa, *Intellectual Property Rights, the WTO and Developing Countries: The TRIPS Agreement and Policy Options*

Bhagirath Lal Das, *An Introduction To The WTO Agreements*

Bhagirath Lal Das, *The WTO Agreements: Deficiencies, Imbalances and Required Changes*

Bhagirath Lal Das, *The World Trade Organization: A Guide to the New Framework for International Trade*

Biplab Dasgupta, *Structural Adjustment, Global Trade and the New Political Economy of Development*

Graham Dunkley, *The Free Trade Adventure: The WTO, the Uruguay Round and Globalism: A Critique*

S. R. Khan (ed), *Trade and Environment: Difficult Policy Choices at the Interface*

John Madeley, *Hungry for Trade: How the Poor Pay for Free Trade*

Hans-Peter Martin and Harald Schumann, *The Global Trap: Globalization and the Assault on Prosperity and Democracy*

Vandana Shiva, *Protect or Plunder? Understanding Intellectual Property Rights*

Oscar Ugarteche, *The False Dilemma: Globalization – Opportunity or Threat?*

In preparation

Carlos Correa and Nagesh Kumar, *Establishing International Rules for Foreign Investment: Trade-Related Investment Measures (TRIMS) and Developing Countries*

Caroline Dommen, *Trading Rights? Human Rights and the WTO*

Graham Dunkley, *Free Trade Mythologies: Trade, Globalization and Development: Critique and Alternatives*

Ha-Joon Chang and Ilene Grabel, *There is No Alternative? Myths and Realities about Development Policy Alternatives*

For full details of this list and Zed's other subject and general catalogues, please write to:

The Marketing Department, Zed Books, 7 Cynthia Street, London N1 9JF, UK or email Sales@zedbooks.demon.co.uk

Visit our website at: http://www.zedbooks.demon.co.uk